Digital Technologies and the Museum Experience

Digital Technologies and the Museum Experience

Handheld Guides and Other Media

EDITED BY
LOÏC TALLON AND KEVIN WALKER

ALTAMIRA
PRESS

A Division of
ROWMAN & LITTLEFIELD PUBLISHERS, INC.
Lanham • New York • Toronto • Plymouth, UK

ALTAMIRA PRESS
A division of Rowman & Littlefield Publishers, Inc.
A wholly owned subsidary of The Rowman & Littlefield Publishing Group, Inc.
4501 Forbes Boulevard, Suite 200
Lanham, MD 20706
www.altamirapress.com

Estover Road
Plymouth PL6 7PY
United Kingdom

British Library Cataloguing in Publication Information Available

Library of Congress Cataloguing-in-Publication Data

Tallon, Loïc, 1981-
  Digital technologies and the museum experience: handheld guides and other media /
Loïc Tallon and Kevin Walker.
     p. cm.
  Includes bibliographical references and index.
  ISBN-13: 978-0-7591-1119-6 (cloth : alk. paper)
  ISBN-10: 0-7591-1119-7 (cloth : alk. paper)
  ISBN-13: 978-0-7591-1121-9 (pbk. : alk. paper)
  ISBN-10: 0-7591-1121-9 (pbk. : alk. paper)
  ISBN-13: 978-0-7591-1237-7 (electronic)
  ISBN-10: 0-7591-1237-1 (electronic)
  1. Museums—Technological innovations. 2. Museums—Educational aspects. 3. Digital
media. 4. Pocket computers. 5. Cellular telephones. 6. Digital cameras. 7. MP3 players.
8. Museum visitors—Education. 9. Museum exhibits. I. Walker, Kevin, 1966- II. Title.
  AM7.T355 2008
  069—dc22
                                                                                    2008011394

Printed in the United States of America

♾™ The paper used in this publication meets the minimum requirements of American
National Standard for Information Sciences—Permanence of Paper for Printed Library
Materials, ANSI/NISO Z39.48-1992.

Contents

Acknowledgments   vii

Foreword   ix
    James M. Bradburne

Introduction: Mobile, Digital, and Personal   xiii
    Loïc Tallon

PART I: **Defining the Context: Three Perspectives**

  1   The Exploded Museum   3
    Peter Samis

  2   Enhancing Visitor Interaction and Learning with
    Mobile Technologies   19
    John H. Falk and Lynn D. Dierking

  3   Designing Mobile Digital Experiences   35
    Ben Gammon and Alexandra Burch

PART II: **Delivering Potential**

  4   Audibly Engaged: Talking the Walk   63
    Jeffrey K. Smith and Pablo P. L. Tinio

  5   Mobile Multimedia: Reflections from Ten Years of Practice 79
    Silvia Filippini-Fantoni and Jonathan P. Bowen

6 Improving Visitor Access   97
       Ellen Giusti

7 Structuring Visitor Participation   109
       Kevin Walker

8 Designing for Mobile Visitor Engagement   125
       Sherry Hsi

9 Cross-Context Learning   147
       Paul Rudman, Mike Sharples, Peter Lonsdale, Giasemi Vavoula,
       Julia Meek

10 Interactive Adventures   167
        Halina Gottlieb

11 Afterword: The Future in Our Hands? Putting Potential
        into Practice   179
        Ross Parry

Bibliography   195

Index   223

About the Contributors   231

Acknowledgments

This book has been a distributed effort with contributions separated by time and great distances. We truly believe it reflects the state of the art in terms of theory and technology, and so, first and foremost, we thank all of our authors. Your work will make this a great work for years to come.

Thanks go to those who contributed by reading early drafts and offering ideas, advice, or encouragement: Allison Walker, Giles Waterfield, Pam Meecham, Steve Dale, David Anderson, Liam Bannon, Jane Burton, Margaret Greeves, Victor Kaptelinin, Sara Knelman, Zsuzsanna Kondor, Richard Noss, Donald Peterson, Palmyre Pierroux, Kathryn Potts, Dagny Stuedahl, Chris Tellis, Jenny Uglow, Niall Winters.

Finally, a word of thanks to those whose work or ideas provided inspiration or guidance, either directly or indirectly, including: Nicolas Balacheff, Robert Barnett, Martyn Best, Sarah Bird, Elizabeth Brusca, Toni Bryan, Red Burns, Brian Cantor, Tara Chittenden, Ann Curtis, Robert Cutler, Chris Dennett, Pierre Dillenbourg, Gwen Farrelly, Tim Goalen, Tom Goldstein, Lawrence Grauman, Todd Gitlin, Alan Greenberg, Phyllis Hecht, Erin Hersher, Eilean Hooper-Greenhill, Maht Hussain, Sue Johnson, Cathy Klimaszewski, Paul Lebrecque, Sophie Langdale, Chris Langdale, Diana Laurillard, Mark Levene, Chris Littlewood, Rose Luckin, Barbara Mathe, Theano Moussouri, Laura

Nader, Dan Phillips, Vincent Puig, Fred Ritchin, George Roussos, Vincent Sarich, Jeffrey Shaw, Elvin Sledge, Eric Steinberg, Jean-Pierre Tallon, Susan Tallon, Chloe Thibault, Luke Uglow, Josh Underwood, Shawn Van Every, Madeline Walker, Sherry Wasserman, Willard Whitson, Michele Willens, Ed Woodard.

Foreword

James M. Bradburne

Digital Technologies and the Museum Experience comes at an opportune moment in the history of informal learning. Now, more than ever before, new technologies allow the museum to imagine creating new experiences and enhancing familiar ones in unprecedented ways. *Digital Technologies and the Museum Experience* explores the ways in which mobile devices and digital technology can be used to enhance and transform the visitor's experience of the museum, and looks at the technologies that can extend the museum's ability to invest the world with meaning beyond its own walls, by inviting visitors themselves to contribute to the museum's meaning-making activity.

Traditionally, museums have used written texts as the main instrument to convey meaning—particularly cognitive meaning. At the same time, research demonstrates convincingly that, while no one reads all texts in a museum, nearly every museum visitor reads some texts, in some contexts. The written or spoken text remains one of the most efficient ways for humans to learn new facts and stories, and it is ubiquitous in museums and similar institutions. From the moment visitors arrive, the museum (its designers, its educators, and its curators) is constantly placing constraints on their experience, as Jeffrey Smith, Pablo Tinio, and Kevin Walker discuss in this book. In fact, there are relatively few unconstructed and unmediated moments in a museum setting; the museum wittingly or unwittingly shapes virtually every aspect of the visit. For instance, we may decide to put a Picasso painting beside an African mask. By doing so we have deliberately made meaning and are inviting the visitor to explore that meaning. We often think that what we are communicating is purely content. But the meaning we want the user to construct is not just about the relationship between Picasso and his sources; it is also about the relationship between the museum and the visitor. Who is in control? Who makes the meaning? Who gets to say which features were important to Picasso and which weren't?

Meaning making in the museum has always had a social and political dimension. Willem Sandberg, director of the Stedelijk Museum from 1945 to 1962, who pioneered the first museum audio tours, also pioneered unjustified text (flush left, with equal word spacing), which he believed challenged convention and had important social overtones. Sandberg was among the first to recognize the importance of the visitor's as well as the museum's voice, and to argue that they consisted of a dialogue, not only a "top-down" lecture. Along with Marshall McLuhan, Sandberg was among the first to champion the ways in which the museum had to transform itself—long before the technology was available to do so. In the last decade, technological advances have fostered innovative ways of delivering text and have challenged museums to rethink traditional ideas about what text-based interpretation is and can be. Technology has also had an impact on visitors' expectations about how they will receive—and contribute—information in the museum context. Communication tools such as the iPod and Web-enabled mobile phones, which let users augment gallery visits with off-site "unauthorized" video and audio content, mean the museum spaces are being opened—willingly or not—to voices other than those of the curators. These days, the motivated visitor can arguably reconfigure a gallery visit to meet his or her own specific needs—with or without the museum's help.

So when we speak of digital voices in the museum, on the one hand we mean the museum's subtexts and the museum's social contexts—in short, the sum of all the elements that make the museum a meaning-rich environment for its users. On the other hand, the book signals an important new dimension in museum interpretation. For years, the space of the museum has been the preserve of curators and educators, who were solely responsible for the museum's content. In recent decades there has been an increasing insistence on "bottom-up" approaches that open the museum to other voices and other constituencies. The idea of visitors contributing to the museum space is not new, nor is it dependent on digital technologies—after all, as the "Memory" exhibition at the San Francisco Exploratorium showed over a decade ago, sticky notes and a box of pencils is enough to invite a community to rewrite its own history. Nevertheless, the suggestion that visitors can—and should—contribute to the museum environment is both radical and provocative.

But do visitors "get" the texts, whether displayed on a screen, written on the wall, or digitally whispered in their ear? Which texts do they understand and in what ways? What does technology add to, or take away from, the meaning-making capacity of the museum? In a media-rich modern museum, in what ways will text be transformed as it is used and enhanced by increasingly powerful devices? The museum ideal is visitor engagement—but this means more than the self-sustained activity of a hamster on a treadmill; it is self-absorbed concentration in which users direct their own learning. Coincidentally perhaps, this is the behavior that characterizes puzzle solving, working in an interactive laboratory, playing a computer game, or reading a murder mystery. Self-initiated, self-directed, self-sustaining engagement is a hallmark of experiences whereby we learn, in Jonathan Miller's words, that "the life of the mind is a pleasure."[1]

Whether as top-down information or bottom-up dialogue, it is important to understand the ways in which the relationship in the museum is structured. So, in order to speak of ways in which we can alter—and enhance—the quality of visitor engagement, I would like to introduce a new term, borrowed from systems research: that of the "user language." As defined by Dutch theorist Gerard de Zeeuw, a user language is the "collection of constraints that helps shape the variation generated by an actor into patterned behavior."[2] The means through which we structure a specific relationship to the visitor by means of our exhibits is the user language. The exhibit's user language describes what counts: what can be included and what is irrelevant. Different user languages confer specific properties on the user. In the museum the most significant user languages are "textual authority," "observation," "variables," "problems," and "games," and each user language invites the museum visitor to relate to the museum setting in a different way—as a listener, as an observer, as an investigator, or as a player. The notion of a user language allows us to describe and analyze the museum "text" in terms of both its content and its intent.

Seen in this way, the museum "text" not only gives information but also proposes a relationship between the text writer and the text reader. So, it is useful to ask which user languages are best supported by what medium. Are certain technologies better at supporting certain user languages than others? If our goal is to increase engagement in the museum setting by the use of user languages that confer greater agency on the visitor, in what ways can new technological platforms and affordances help us reach that goal? This book looks at the ways in which our experience of digital technologies can help enhance the potential of the museum space—and allow visitors to enter into a dialogue with the museum. By their nature, digital technologies offer visitors the opportunity to contribute, affect, and potentially subvert the meaning-making enterprise of the museum. By championing the importance of the digital dialogue, making digital voices heard, and listening to what they have to say, this book makes an important contribution to the field of contemporary museum studies.

NOTES

1. Jonathan Miller to James M. Bradburne, unpublished letter, 1997.

2. Gerard de Zeeuw, *Coordinated Cooperation and Increasing Competence* (Amsterdam: UVA/OOC, 1990), 3.

Introduction: Mobile, Digital, and Personal

Loïc Tallon

The first visitor technology used in a museum was handheld. When invented in 1952, the developers then, like developers today, were drawn by its unique potential to mediate an experience individually controllable by each visitor, which was content rich, was personal to them, was available at any time, and suited learning styles not served by catalog, text panel, or label.

That said, the first museum handheld technology, the Stedelijk Museum's Short-Wave Ambulatory Lectures, barely delivered on this potential. This was not due to a lack of vision by the developers; they recognized the medium's potential, enthusing, "The possibilities of this device are so great that in the future shortwave lectures cannot be ignored in any museum. In the future shortwave lectures discussing individual works of art will be installed in such a way that they can be heard by any visitor at will."[1]

The analog technologies of the 1950s did not have the capacity to fulfill this vision. Ambulatory Lectures were delivered through a closed-circuit short-wave radio broadcasting system in which the amplified audio output of an analog playback tape recorder served as a broadcast station, and transmission was via a loop aerial fixed around the gallery or galleries. Identical lectures in Dutch, French, English, and German were recorded onto magnetic tapes, broadcast in turn through the aerial, and picked up by visitors through a portable radio receiver with headphones, when inside the loop. A technological achievement in itself, the system was such that all visitors with a receiver could only hear a specific piece of commentary at any time; hence, groups of visitors would move through the galleries and look at exhibits as if guided by an invisible force, in complete synchronicity.

Developments in hardware, content creation, and functionality have since enabled ever more powerful handheld guides to deliver better on the medium's unique potential. From its origin as an analog radio tour at the Stedelijk Museum, through its use by over three million North Americans as a Sony Walkman-style taped tour of the eight-stop "Treasures of Tutankhamun" exhibitions in the late 1970s, to its incorporation as a direct-access—also known as random access—digital guide to the Louvre's permanent collection in 1993, to its subsequent adoption by virtually every major museum by the end of the twentieth century, and to its establishment at the forefront of in-gallery interpretation innovation, handheld technology is today an established companion of the modern museum. Handheld guides today exist in sites ranging from the Forbidden City to the Empire State Building, and from the London Zoo to the San Francisco Museum of Modern Art (SFMOMA). It is telling that, until the Internet, handheld technologies—and specifically audio guides—were the only visitor technology to have been universally adopted by museums.

DEFINING POTENTIAL

Understanding the handheld guide's long and wide application in museums today is not simply about grasping its unique potential; a jukebox has unique potential, but there are few good reasons to install one in a museum. More fundamental—and true for any visitor provision, digital or otherwise—is an understanding of visitors' needs and expectations. Only with this understanding can the unique potential of handheld technologies be harnessed to meet these visitor expectations and thus deliver a more rewarding museum experience.

The trend is toward personal relevance and interpretations, interactivity, and easy access and control of content to shape the twenty-first-century museum visitor's experience. Today's museum visitors are less audience than they are author—active participants in meaning making and content creation.[2] Where once radio show phone-ins and letters to newspapers were the main means by which members of the public could voice their opinions, today the most popular television shows are "reality" shows where viewers can vote for

A Sound-Trek audio guide in use at the American Museum of Natural History, New York City, in 1961. As a portable radio-receiver operating within a closed-circuit broadcast, the dial on the front of the device allowed visitors to literally tune in to the commentary audio narration of their choice.

A Telesonic Lorgnette "radio guide" in use at the Science Museum, London, in 1961.

their preferred outcomes, and many of the most popular Internet sites are those specifically conceived for user-led content creation and knowledge sharing. And there is the "long tail"—active dialogues exist for virtually all specializations and interests.[3]

Layered over this information and content network is the increasing expectation to be able to access and enter into these dialogues whenever one wants, anywhere and at any time. In its simplest form this is being able to talk to a friend at any time, no matter where in the world they are. But it also means finding the snow conditions in the Alps from the car on the way to the airport, or following the progress of a NASA rover in its exploration of the surface of Mars. Mobility and connectivity are fundamental expectations.[4]

To remain relevant to today's public, museums must follow these trends and meet these evolving visitor expectations. The question is how best to do so.

When the Stedelijk Museum was addressing the needs of foreign language visitors but intent on maintaining the traditional "sacred silence" of the galleries, it turned to a combination of radio and cassette-tape technology. For the Tutankhamun exhibitions, the original blockbuster, American museums turned to a Walkman-styled technology solution in order to engage new audiences unaccustomed to the traditional didactic tools of an exhibition. When in 1998 the Experience Music Project in Seattle wanted to deliver a personal interactive experience covering the entire museum visit, it turned to multimedia handheld guides. When Tate Modern sought to augment their provision for deaf visitors, it experimented with sign-language tours on a personal digital assistant (PDA). And when SFMOMA wanted to extend its experience beyond its walls and personally engage with potential visitors in their everyday lives, it developed a monthly podcast. These examples are not "proof" that technology is the only means of meeting visitors' expectations; that is far from the case. However, the fact that museums have time and again turned to technology to meet visitor expectations is due to technologies' ability to deliver new interactions and experiences. If they didn't, there would be no need for technology in museums.

The versatility of modern digital technologies in managing and delivering large quantities of information to suit each visitor's preferred learning style and approach is central to its unique potential. As new technologies evolve and wireless Internet access becomes more stable and ubiquitous, the potential of digital technologies will only increase, making ever more powerful new experiences a reality. And when this potential is combined with that of mobility, there evolves a uniquely strong capacity to meet modern visitors' expectations.

Carried by the visitor, by their very essence handheld technologies have a one-to-one personal relationship with the user: the user has personal control over how to use them and the content they contain. Furthermore, with PDAs, mobile phones, portable media players, and digital cameras are converging handheld technologies already in the hands of a wide public, comfortable and literate with their modes of engagement. These are technologies with which users have a ready-made, intuitive relationship—one museums could tap into. And technologies owned by visitors can provide particularly cost-effective solutions, accessible to museums of all types and sizes. Simply stated, handheld digital technologies have the potential to mediate personally rewarding museum experiences that no other medium can replicate.

MOBILE, DIGITAL, AND PERSONAL

Handheld technologies in museums have gone from being one technology—a portable audio player or audio guide—to being many, each with specific functionality. No single term exists that categorizes these now disparate handheld technologies. Since their invention, audio guides alone have been called anything from guide-a-phones to walkie-talkie tours, from acoustic guides to Acoustiguides. Multimedia tours—also referred to as handheld or PDA guides—and consumer technologies such as mobile phones, digital cameras, and MP3 players today fall into the same museum category as audio guides.

Mobile, digital, and personal: these are the three defining qualities of this technology category. They are mobile in that they are location independent, available anytime, anywhere; they are digital in that their functionality is based on an electrical system that uses discrete values; and they are personal in that there is a one-to-one relationship between the visitor and the medium, with the visitor in control.

These qualities also highlight the museum context within which handheld guides exist. With only digital and personal as defining qualities, multimedia kiosks would fall within the categorization. Lose digital too and you have the label and wall text. With only personal and mobile but not digital, you have the museum catalog. And with only mobile, you have the docent or tour guide, the original "interpretive aid" with a history dating back as far as Giorgio Vasari, J. J. Winckleman, and the origins of the art history discipline and the museum.

THE RESEARCH CONTEXT

For a medium with such an extensive existing usage and wide-reaching potential, there is a distinct lack of rigorous, accessible, and published research. In the early years of handheld guide development, museums undertook a series of practical research projects with third-party hardware suppliers. Naturally, the research was hardware focused, the objective being to develop a durable, minimal-maintenance, and cost-effective hardware solution for the mobile delivery of audio. Research into the visitor experience generated by the audio was cursory, often based solely on take-up statistics—the proportion of visitors who opt to use the handheld guide—and time spent in galleries. After lengthy experimentation with radio wave-based handheld guides—similar to the Stedelijk's—and a brief foray into portable record players, the purpose-built portable cassette player became the hardware of choice. Those created by Acoustiguide, the earliest developers of such a solution, were the industry standard—so much so that, in North America, Acoustiguide is eponymous. However, once a stable hardware solution was developed, research into handheld guides virtually stopped. Only now are museums reacting to this forty-year research silence.

That hardware provision and, more particularly, content production was outsourced by museums to commercial companies instead of developed in-house is central to this research silence. Research creates knowledge, but for any commercial company operating in a competitive market, knowledge is an asset; to share that knowledge, or "expertise," is to dilute one's assets and competitive edge, and thus benefit competitors. With audio guides, Acoustiguide's market leadership was unsuccessfully challenged by By-Word—a company Acoustiguide eventually bought—and then in a more sustained manner by Antenna, known today as Antenna Audio. Since the late 1980s, Acoustiguide and Antenna Audio have held a duopoly on the museum handheld technology market: with the two companies consistently competing for the same museum contracts, a particularly tight commercial rivalry has evolved. Inevitably, therefore, what research Acoustiguide and Antenna Audio have undertaken has been rarely disseminated, and that which has is invariably measured by commercial interests.

But this is just half the story; as clients, museums are equally culpable for the research silence. At major museums, financial incentives rather than curatorial purpose have dictated, and still do dictate, that audio tours be installed. Since the 1970s the common business relationship between handheld guide provider and museum has meant that the former develops content and provides the hardware at no cost to the museum, but that the museum receives a share of profits generated. For many major museums, handheld guides are a successful revenue generator and, better still, one that requires minimal risk or investment. Though it is difficult to obtain the exact figure or share of the revenues—these are kept confidential, often covered by nondisclosure agreements—gross audio tour revenues from the aforementioned Tutankhamun exhibitions topped five million dollars. Money speaks louder than research.

Counterposing the museum's focus on financial benefits, there has also existed an overt intellectual prejudice against handheld technologies (the past tense here reflects my optimism rather than proven realism). It is not uncommon for a curator, at the opening of an exhibition they have spent months, if not years, preparing, to have not listened to the accompanying audio tour. Is it not strange that despite applying studious and tight control over the minutiae of what visitors will see—from the exhibition layout to bench positions, and from the display cases to the label content—only curatorial lip service is paid to what a significant proportion of visitors will carry with them, hear, watch, and interact with?

From the medium's invention, the museum industry has held deep reservations about the "digital dialogue" that handheld guides mediate. As a 1960s *Museums Journal* review noted,

> It is a fact beyond doubt that a great many visitors like to wander at will, stand and stare, and equally dislike any breath of regimentation. There is a danger that with the wide application of mechanical gadgets the quality of visitors may suffer. There are many who would be dismayed if they saw throughout the building people with black boxes around their necks pass by with a faraway expression in their eyes . . . guided by some mysterious forces they walk, turn, and stop in almost synchronized precision before exhibit after exhibit.[5]

Handheld technology users have been accused of being antisocial, of clogging up a gallery, of only being interested in exhibits covered by the guide, and of only engaging passively with artworks. Screen-based multimedia tours are further accused of distracting the visitor from looking at the exhibit; but do visitors look at exhibits while reading labels?

Such criticism personifies handheld technologies as a master rather than a tool. One UK art critic proclaimed that the audio guide is "a ruse to squeeze an extra few quid from gullible patrons happy to amble around like zombies while a disembodied academic voice tells them what to think."[6]

Intelligent criticism of handheld technologies focuses on the nature of the viewer-exhibit interaction it mediates. Qualifying this interaction is an ideological enterprise informed by a series of assumptions regarding how someone should interact with an artwork. Within contemporary education discourses, the key concern is that the visitor receives a personally rewarding experience.

In fifty years, handheld technologies have gone from simply offering custom content on a standardized device with set functionality—the audio guide—to offering custom content on different hardware platforms with custom functionality suited first to the museum but increasingly also to the visitor. Today, the hardware platform for a handheld guide can be purchased off the shelf, and limited specialist technical knowledge is required for either programming the hardware or developing and producing the content. This has opened up the handheld technology museum market.

Today, practical research exploring handheld technologies and mobile learning is carried out in universities—institutions in which knowledge sharing is fundamental. Such research explores the functionality and content styles of handheld technologies as a means of understanding and further enhancing the visitor experience. Museums are now sharing and participating in this activity and knowledge.

Furthermore, it is understood that a handheld guide, just like any other visitor service, cannot be all things to all people. Different people respond to different stimuli in different ways: what works for one museum visitor may not work for the next. It is about creating a variety of different portals through which to engage with an exhibit. This has focused researchers on the interactions that take place, be they mediated through an audio guide, digital camera, MP3 player, mobile phone, or PDA. It is on the fundamental issues that underpin these different interactions that *Digital Technologies and the Museum Experience* focuses.

OBJECTIVES OF THIS BOOK

This book is not about technology per se; it is about visitors' dialogue and interaction with cultural stimuli in museums, and how handheld technologies can be used to enrich that dialogue and deliver more rewarding museum experiences. By analyzing and comparing experiences from a wide range of museum projects across the globe, this book seeks to define the social context and learning framework within which successive generations of handheld guides should be developed. *Digital Technologies and the Museum Experience* does not attempt to predict the future of handheld guides; it strives to help the museum community invent it.

As such, the definition of "museum" that contributors have adopted is inclusive. It includes art galleries, science centers, zoos, aquaria, botanic gardens, heritage sites, and museums alike. These various informal learning centers differ in the nature and function of what they offer visitors, and these differences affect how handheld technologies could be successfully applied. However, while the way they are applied will be particular to specific museums, the concerns and issues that shape how they are applied are shared across all types of informal learning centers. One rule does not fit all, but this is less a book about rules than about the issues that lie behind them—issues of mobile learning, inclusion, access, interactivity, personalization, and ultimately, the creation of a more rewarding visitor experience.

To this end, *Digital Technologies and the Museum Experience* consciously avoids focusing on specific technologies, how those technologies work, or what they make possible. The rate of technological change is such that, as with the Stedelijk radio guide, today's impossibility is tomorrow's reality. Any attempt to analyze a temporal snapshot of these technologies would be out of date within months. Instead, and more pointedly we believe, this book focuses on the fundamental visitor and museum issues that shape how handheld guides have been, are, and will continue to be used by museums and visitors. For while digital technologies will continually change, with the technological barriers of today being overcome tomorrow, the issues that shape how handheld guides have evolved date back to the origin of such guides and have remained consistent.

For example, increasing visitor access prompted the Stedelijk Museum's innovation and the use of audio tours for the Tutankhamun exhibitions. Using the medium to customize and personalize the visitor experience was first

A dual head-phoned Acoustiguide audio guide in use at the National Gallery of Art, Washington D.C., ca.1965. These portable cassette players would take over from radio guides as the dominant audio guide system, so much so that today in North America, Acoustiguide is eponymous.

sought in the 1960s at the Smithsonian Museum of Natural History, by providing visitors with a variety of tour styles to chose from, all accessible on the same device. Experiments with interactivity on handheld guides began in 1973 at the Milwaukee Public Museum. This included audio narration, played on a portable cassette player, that automatically stopped after asking a question, only resuming once the visitor had responded by punching a correct answer into the attached punch card reader.[7]

Attempts to develop a more social handheld guide experience also originated in the 1960s with audio guides that included headphone sockets for two listeners: at the National Gallery of Art in Washington these were more popular than their single-listener counterparts. Structured learning trails were inherent to early linear taped tours, where each consecutive audio stop could build upon the ideas discussed in previous audio stops. Arguably, learning trails only became an overt concern for handheld guides when the invention of direct digital access gave visitors control over the order in which to listen to audio, and so a structured buildup of knowledge could not be developed into the audio narration.

The solicitation and promotion of the visitor's voice, as opposed to the museum's authority, was sown in 1985 when Creative Time in New York developed a little-known audio tour of the Metropolitan Museum entitled "Masterpieces without the Director." Spliced together from anecdotal interviews with visitors in front of specific artworks, the tour challenged the museum's authority. For example, when looking at the ca. 2051–2000 BCE statue of Mentuhotep II, comments included, "It's so alive and it's so full of . . . [pause] it's perfect," and, "In fact every museum has one of these, and to be brutally honest, it just reminds me of those found in sushi bars." Instead of a beep to signal the end of audio segments, there was the sound of a starting gun! Produced by a theater group and thus not an official Met audio tour, it required visitors to use their own personal cassette player to listen to the tour, which could be procured from vendors on the museum steps. So, as well as challenging the museum's authority, "Masterpieces without the Director" simultaneously initiated the use of visitors' own hardware for handheld guides, a trend that continues today with MP3 players and mobile phones.

Even contemporary concerns with extending the museum experience beyond its walls can be traced back to over half a century ago: in the 1950s the American Museum of Natural History would broadcast regular lectures on public radio, at the end of which listeners were invited to visit the museum at a set time for an accompanied tour of the exhibits discussed during the broadcast.

Using past experience as its guide, *Digital Technologies and the Museum Experience* explores a wide range of cutting- and leading-edge research and implementation projects involving handheld technologies from across the globe, uniting these strands into a common thread for the first time. We hope the theoretical and practical issues, the design principles, and the evaluation methods shared herein will remain relevant for many years to come.

NOTES

1. Erich Wiese, "Experiences with Short Wave Radio Tours in the Hesse Museum at Darmstadt, in Museumskunde" (unpublished letter by Der Deutsche Museumbund, Verlag Walter de Gruyter & Co., Berlin, 1960/1961).

2. Andrew Keen, *The Cult of the Amateur: How Today's Internet Is Killing Our Culture and Assaulting Our Economy* (London: Nicholas Brealey Publishing, 2007).

3. Chris Anderson, *The Long Tail: How Endless Choice Is Creating Unlimited Demand* (London: Random House Business Books, 2006).

4. Howard Rheingold, *Smart Mobs: The Next Social Revolution* (Cambridge, Mass.: Basic Books, 2002).

5. "Editorial," *Museums Journal* 5, no. 60 (August 1960): 112.

6. Alfred Hickling, "Block Beuys," *Guardian*, November 29, 2004.

7. Chandler Screven, *The Measurement and Facilitation of Learning in the Museum Environment: An Experimental Analysis* (Washington, D.C.: Smithsonian Institute Press, 1974).

I

DEFINING THE CONTEXT: THREE PERSPECTIVES

1

The Exploded Museum

PETER SAMIS

In a technological world, the museum visit no longer begins when a person enters the building, nor need it end when she or he leaves.[1] The museum's physical space is but one site—albeit a privileged one—in the continuum of the visitor's imaginative universe.

Contemporary artist Olafur Eliasson challenges the traditional model of the museum encounter:

> The very basic belief behind my work is that objecthood, or objects as such, don't have a place in the world if there's no subjectivity, if there's not an individual person making some use of that object. This even goes for gold and diamonds . . . [and] within art, it's even more mystifying, it's more mystical, it's even more alienating. The objecthood is money in the bank, regardless of whether people look at it or use it or have it around them or not. Because this is very counterproductive to what I think is essentially important: namely, that individuality and the nature of individuality . . . has to be reconsidered constantly, as a model, in order to sustain itself in the world today, in order to have an impact on the world today. If the object becomes prescriptive of the individual, of the subject, then we don't integrate time as time passes along. . . .
>
> I think there is a paradox, looking at the history of museums . . . collecting objects from reality, preserving them in a container somewhat outside of reality. . . . Museums today, in my view at least, should be a part of the world, a part of the times in which we live. Even if they have historical collections, they still need to emphasize the fact that you are looking at them from where we are today.[2]

Back in 1966, Bob Dylan wrote,

> Inside the museums, Infinity goes up on trial
> Voices echo this is what salvation must be like after a while[3]

Dylan's line was a prescient observation, a precursor to the plethora of publications that, starting in the late 1980s, critiqued the museum field's absolutist stance and protection of Western cultural norms.[4] Indeed, in the intervening decades most museums have continued to rely on the authority of their presentations, as if there were one set of objective truths to be gleaned from the objects in their custody, and the visiting public was an undifferentiated set of empty vessels to be ignored or filled with *la science infuse*.[5]

Meanwhile, some artists, curators, and educators did try to poke at the sacrosanct boundaries of museum practice, from both inside and outside the field. Howard Gardner's "multiple intelligences" theory exploded the "one size fits all" model of education,[6] while constructivist learning models emphasized that no one comes to museums as a clean slate, an empty vessel waiting to be filled. In addition to learning styles and aptitudes, each person brings with them a personal history, a psychobiography, and engages the museum within a social context, visiting alone, with friends or associates, or with family.[7]

Taken together with increasingly granular audience segmentation research, the visitor-studies literature on museums has boomed over the past ten years, and we are coming to know our "guests" in an increasingly sophisticated way.[8] In fact, more and more museums are professing an interest in what that cipherlike presence, the once-anonymous visitor, has to say. So while Dylan's next line—"But Mona Lisa musta had the highway blues / You can tell by the way she smiles"—would not have been out of place as a subjective visitor observation in the past, and might have found its way into an inquiry-based docent tour, now it might be published on a museum blog!

Eliasson's emphasis on the perceiving subject echoes the contemporary preoccupation with the varieties of individual experience. The museum as a commodifying factor, a temple on high, is dethroned, and the visitor, with whom all experience must finally succeed or fail, thrive or fall on barren ground, is deemed the final arbiter. The museum is the sum not of the objects it contains but rather of the experiences it triggers. To quote San Francisco

Museum of Modern Art (SFMOMA) senior curator of painting and sculpture Madeleine Grynsztejn,

> We don't do our best when we simply instruct. We do our best when we answer questions alongside the visitor, and ask questions alongside the visitor. And when we create a kind of conversation . . . . We don't do our best when we create a one-way dialogue that is assertive and one-dimensional. We do our best when we offer multiple avenues of interpretation, and when we keep a lot of room for audience response.[9]

EXPLODING INTERPRETATION, IN PRACTICE

So how has the use of interpretive technologies paralleled this shift?

We all know: in the beginning was the wall text. And whether it was good or not, it was all visitors had, and we came to depend on it. I use the word "we" advisedly, to include both professional museum staff and our visiting public. The idea of frontloading all the essential concepts for the appreciation of a long and complex exhibition before a visitor has seen a single object is inherently flawed, and yet many museums continue to do so.[10]

Next came extended object labels, in which basic artist/title/date info was augmented by a paragraph of text, and with them an acknowledgment that visitors might require additional information "just-in-time," as they stand in front of an object. (Believe it or not, even today, many exhibitions still do not include them.) Both wall texts and object labels have typically been monovocal, written in an anonymous and authoritative "museum voice."

Corresponding to these typeset texts was the linear audio tour, which had the virtue of channeling the museum voice into your ears as you stood in front of an object, thereby liberating your eyes to actually see it. The "empty vessel" model of knowledge acquisition was alive and well, and our ears were the apertures best fitted for filling.

In the late 1980s to early 1990s, two innovations occurred—one due to a change in the audio tour business, the other to the advent of new technologies. The first innovation was philosophical: the master narrative as promulgated by a single authoritative museum voice gave way to a polyphony of voices, and with them the admission of more than one perspective in evoking the value and meanings of a work of art.[11] The second innovation was digital:

the ability to randomly access as much or as little information as you wanted about an object in the gallery, and to pick and choose your way through an exhibition without the museum determining your course. (Exhibitions, of course, remained linear, unfolding in space, but which objects you chose to commune with, and which tour stops to consult, suddenly became your call.) Several other chapters in this book discuss linear versus random-access audio tours.

Taken together, these changes were big; the monopoly of the expert was challenged. Not that people didn't want to hear an expert; by most accounts they do. But the multiplication of points of view pointed to the "many meanings all happening at once" nature of the world, and showed that museum objects are no exception. It meant that multiple entry points could be equally valid for experiencing art and artifacts, meshing with the learning styles and entrance narratives of a variety of visitors.[12] And it turns out that that is one of the things Web 2.0 is all about.

TALKING AND TAGGING

On May 28, 2005, the *New York Times* reported in a front-page story that a professor and his students at Marymount College in Manhattan, dubbing themselves "Art Mobs," had brought their digital recorders into the galleries of New York's Museum of Modern Art (MoMA) and created a set of guerrilla podcasts—alternative audio perspectives on some of the major works in MoMA's permanent collection. These commentaries were available as free downloads on the Web.[13] The news rippled like shock waves from an earth tremor in the museum world. For the first time (or rather the first time again—as artists have a long-standing tradition of undermining museum authority going back at least as far as Marcel Duchamp), someone had publicly usurped the museum voice from an esteemed, authoritative institution and substituted a set of opinionated, perceptive, and irreverent alternatives. Canonical works were no longer hallowed; in fact, some were actively ridiculed. The critics had taken to the airwaves and invited listeners to take their voices along on their next visit to the museum.

MP3 players and the advent of podcasting empowered members of the public to publish their own perspectives and stories on subjects as far flung as anime and politics. Museums were but one of thousands of potential topics, but we weren't used to having anyone else occupy our territory. We weren't used to having to share our space.

MoMA replied by posting their entire audio tour online for free download (in fact, the museum had coincidentally just secured outside funding to make this possible) and by inviting potential visitors to create their own audio programs at home for personal use on their next visit—arguably a rather labor-intensive proposition for a limited audience of one or two. They did not actively solicit public submissions, nor did they exclude the possibility of considering them.[14] But all things considered, their response was rather enlightened: most museums still charge for their audio tours and do not yet post them to their websites. The net effect: free distribution of the master narrative beyond the museum's walls, which, when paired with an increasingly complete online representation of the collection, makes for an informative virtual visit.

Other museums took different approaches with their podcasts. At the San Francisco Museum of Modern Art, we created SFMOMA Artcasts, an online, illustrated audio zine designed to project a variety of art concepts and voices out into the community, and to invite the community back into the museum.[15]

The voices on the Artcasts include the artists themselves (one of the distinct advantages of being a museum of modern and contemporary art), curators, "Guest Takes" where poets, composers, and musicians are invited to respond in their own art form to works on view, and "Vox Pop," where nonexpert members of the visiting public are asked to reflect on what they're seeing in the galleries. The mix of voices and genres creates a lively dialogue, while our collaboration with Antenna Audio preserves the production values traditionally associated with the museum.

Many other museums now also produce podcasts for various constituencies and with varying degrees of finish. Younger audiences are often targeted, as with Tate Modern's "Raw Canvas" and MoMA's "Red Studio" podcasts, produced by and for teens in collaboration with professional sound designers/engineers. In all cases, the museum retains final editorial control over what it publishes, even as it expands the array of voices and perspectives it presents.[16]

Is there a line museums will not—or should not—cross? How far will we go in accepting visitor contributions to our officially published content? This seems to be the frontier du jour, and it is playing itself out on multiple museum horizons as of this writing. How willing are we to break the prover-bial fourth wall and listen in as our visitors describe what they see in our gal-leries and how they connect art to their lives—or fail to do so? Do we really want to know?

The steve project (www.steve.museum) is one such test case. A research col-laboration developed by an alliance of North American museums and funded by the Institute for Museum and Library Services, it aims to test a number of hypotheses about how user-generated tags might aid in the description of—and facilitate access to—museum collections.

An example will suffice to illustrate the discrepancy between official mu-seum cataloging and the cultural literacy of everyday visitors: one of the his-torical impetuses for the steve project was the realization that a Web search for "Impressionism" on the Metropolitan Museum of Art's collections site would have omitted most of the institution's holdings. All of the paintings corre-sponding to that term were listed in the Met's collections management data-base as "French," "19th century," and "oil on canvas"—but there was no field for "art movement," and hence there were only scattered returns on the word most educated visitors would use to find them. The simple theory behind steve and other efforts at cataloging by crowds is that, if museums use terms submitted by visitors to tag their artworks, other visitors will have an easier time finding them.[17]

Of course, the terms supplied by visitors about an artwork will not all be art related. Many will be subject based, describing the image content of repre-sentational artworks, another frequent lacuna in collections management databases. They will be at the intersection of those artworks and the viewers' vision, and inevitably preconditioned by the viewers' lives (see Falk and Dierk-ing's "personal context" discussed in chapter 2). So, among other questions the steve project will address are the following:

- What kinds of terms are useful to others? What kinds of terms are not?
- Useful to whom?
- How will these terms be validated?
- Will statistical agreement among taggers about a given term be enough to ensure its validity? Or does each term need to be reviewed by museum staff?
- Is such a scenario practical or even possible?
- How will subjective responses—for example, to abstract art—be treated? Are they helpful to others?

All of these questions and more apply as museums enable tagging or commenting on their exhibitions and collections via mobile devices.

A case in point: at the beginning of this chapter we referred to Olafur Eliasson and his critique of the commodifying aspects of museum experience. Eliasson warns of museums' tendency to "package" their messages in a reductive set of bullets fit for public consumption, preempting the visitors' direct perception of the art. In fact, he takes his argument a step further, emphasizing as key to his project the unique and nonrepeatable nature of each visitor's response, and the lack of any one definitive or authoritative experience of an artwork or exhibition.

> Who has the responsibility for seeing what we see? . . . The qualitative potential, let's say, of a work of art lays within the generosity or the sustainability of your engagement. . . . Does it have a potential that adds something to you that you could use in a different context? . . . I would emphasize the importance of looking at the picture as a way of looking at yourself looking at a picture; seeing yourself sensing, or seeing yourself seeing, if you want. . . . So it's not, again, about the museum but about the spectator. . . . So there's something quite generous here, in my view . . . the fact that the museum gives time back to the spectator.[18]

In this context, interpretation can only be an accumulation and juxtaposition of different experiences, none definitive but each building a case for what is commonly held or individually specific.

Admittedly, Eliasson's project is an extreme example, focused as it is on ensuring that each visitor gain a sense of "criticality"—or perspective—on their own experience in the galleries. Furthermore, his works themselves are immersive environments—not so much "on view" as "encompassing you." So we may ask, Do the points Eliasson makes about perception apply equally to paintings hung on a wall, ancient Egyptian statuary, medieval liturgical censers, and Persian calligraphy: Is the dose of dialogue between what the museum and visitor bring to bear a constant, or a sliding scale?

What is the proper titration between expert knowledge and visitor inquiry or response? What is to be gained by facilitating such inquiry, both for the individual and the museum, which stands to increase its knowledge of its visitors and its holdings? To take this line of thought one step further, what might be gained by publishing a range of visitor responses for other visitors, both virtual and physical? How should such responses be organized, so as not to be overwhelming in an information economy already characterized by infoglut?[19]

Museums are just beginning to explore these questions, which, on one level, can be seen as taking the informal conversations that have always taken place between visitors to the galleries, and the more structured dialogues that take place on docent tours, and giving them a published and searchable status on the Web, for public or private consumption. In this light, as technology evolves, new possibilities emerge. For example, as audio tours are delivered over cell phones, the new devices offer visitors the chance to both listen and talk back.

In the United States, a number of museums are beginning to avail themselves of this feature. At the San Jose Museum of Art, visitors were asked to leave comments about the artworks on view in a "Conversation Gallery." While the responses were not as numerous or well considered as they had been in their "Collecting Our Thoughts" exhibition in 2001—where visitors were invited to write wall labels for the artworks with the promise that the best ones would be posted on the gallery walls—there was still sufficient feedback to merit redesigning and repeating the experiment.[20]

It may be that our culture as a whole is becoming more fast paced and oral, and that part of our task is to encourage visitors to slow down enough to, in Eliasson's terms, take their own time. Perhaps we will find that the instant dispatches provided by cell phones are antithetical to such consideration.

MAKING CONNECTIONS

In 1974, Los Angeles artist John Baldessari famously said that, "for there to be progress in TV, the medium must be as neutral as a pencil. Just one more tool in the artists' toolbox, by which we can implement our ideas, our visions, our concerns."[21] It may be said that with the advent of simple editing tools like iMovie and the phenomenal rise of YouTube and other video-sharing websites, Baldessari's prophecy is finally coming true. However, even before YouTube became a household name, museums such as the American Visionary Art Museum and the Denver Art Museum joined places like Grand Central Station in using video booths and storytelling kiosks where visitors could come in, sit down, comment on exhibitions, or add their own memories to the exhibition content. For the most part, these have so far been used in history, children's, and discovery/science museums, and it is perhaps just a matter of time before such video annotation penetrates more art museums as well.[22] One can imagine digital representations of artworks turned into image maps, which can be tagged by multiple viewer users as a common platform for discussion and experience. Such tags could take text or video form, lead to extended annotations, and even include Web links to other far-flung but related sites. In this way, an artwork (or other museum object) can exist both on its own terms and as a hub or focal point for complex interactions—a veritable knowledge interface enabling visitor explorations, associations, and conversations.

Other chapters in this volume describe early work using visitor learning trails and social learning sites such as myartspace.org.uk (now renamed ookl.org.uk). Chapters 7 and 9 describe how students visiting museums were given cell phones to photo-document and audio-annotate their personal itinerary through the museum and respond to a structured set of questions. Once they returned to their classroom, they logged onto the My Art Space site, where they were able to retrieve their captured data and reflect more fully on their experience.[23]

A similar experiment was recently conducted at the Centre Pompidou in Paris, where visitors to an exhibition featuring Iranian photographer and filmmaker Abbas Kiarostami and Spanish filmmaker Victor Erice were able to annotate not only their path through the show but also the entire image track of a film, using sophisticated software called Lignes de temps (Timelines).[24] Walker discusses this in chapter 7.

These efforts to facilitate visitor content creation and publishing represent constructivist learning exercises par excellence. As such, they are probably better suited to the needs and focused time span of a school visit than the harried life of the average museum visitor. Perhaps the "connectivist" paradigm proposed by George Siemens, in which knowing where to find information is as important as having personally made it your own, is more on target for this networked, multiperspectival age in which, in the words of David Weinberger, "Everything is miscellaneous."[25] After all, not everyone is an alpha blogger. In fact, a Forrester Research report suggests that as of April 2007 only 13 percent of those using the Web actively participated by either publishing a blog or a Web page or uploading a video. The vast majority of Web users fall in the less active rungs of this "Hierarchy of Social Participation": 19 percent comment on blogs (the next most active role); 15 percent use Really Simple Syndication

Table 1.1.   Percentage of Barney Exhibition
interpretive offerings used by visitors (n = 251)

Exhibition introduction wall text	78
Exhibition brochure	55
Learning Lounge wall text photos	44
Learning Lounge video	38
Antenna audio guide headset tour	21
Cell phone tour	19
Learning Lounge catalogs	18
Drawing Restraint 9 film	17
Exhibition website	15
Learning Lounge computers	12
Podcast/downloadable tour	7
SFMOMA docent-led public tour	2

(RSS) feeds and tag Web pages; 19 percent use social networking sites; 33 percent read blogs, listen to podcasts, or watch peer-generated video; and 52 percent are listed as "inactives," participating in none of these activities.[26] These figures echo observation of visitor behavior at SFMOMA during the 2006 "Matthew Barney: Drawing Restraint" exhibition:

The sweet spot was clearly on the passive, linear media side—not in the interactive zone. That said, appreciation levels rose most dramatically among those who availed themselves of multiple interpretive resources, including the FAQ wall graphics, artist video, digital audio tour, and interactive kiosk/ website.[27]

Recent research at the Dallas Museum of Art has led to a new model for understanding visitor participation in the art museum. Findings there indicate that, regardless of age, educational level, socioeconomic standing, or ethnic background, visitors fall into one of the following categories:

• Aware: Visitors with little or no experience who are not really comfortable looking at or describing art. They may have been brought to the museum by someone else.
• Curious: Visitors who "like art but are not in love with it." They enjoy the social dimension of museum experience and connections that can be made between art and other parts of their lives.
• Committed: Either educated art consumers who want to be left alone or art enthusiasts who "love art as much as sex and religion" and can't get enough of it, not to mention programs around it.[28]

These levels map nicely to Siemens' first three levels of connectivist engagement:

- Awareness and Receptivity: Learner becomes conscious of new informational nodes/sources of meaning;
- Connection-forming: Learner begins to form connections and uses resources to deepen their knowledge; and
- Contribution and Involvement: Learner contributes to the network, actively gets involved, and becomes visible.[29]

As the visitor progresses in their art experience, the museum promotes personal interpretations over established understandings. In fact, commonly accepted understandings are articulated precisely to open the door to personal responses rather than seal the object in art historical authority.

We are thrown back on the question of the museum-visitor experience and the role of the museum in the visitor's life. Here we are back at Eliasson! We might ask, Is there a continuum of art experience, and where do the museum walls fit within it?

The promise of these new technologies, then, is dual: if they can be made effortless and transparent enough, they can help art ideas to penetrate more effortlessly into visitors' lives, to aid visitors in processing and digesting these ideas and images in their own personal terms. Conversely, new technologies can also open museums to the multiplicity of meanings that our objects trigger in the community of viewers—meanings we haven't yet dreamed of and which stand to be richer and far more diverse than the art historical discourse that is our stock-in-trade.

NOTES

1. This essay is dedicated to the memory of Xavier Perrot, friend and museum colleague extraordinaire.

2. Olafur Eliasson, interview with author, Berlin, June 18, 2007.

3. Bob Dylan, "Visions of Johanna," Copyright ©1966; renewed 1994 Dwarf Music. All rights reserved. International Copyright secured. Reprinted by permission.

4. Among them, Ivan Karp and Steven D. Lavine, eds., *Exhibiting Cultures: The Poetics and Politics of Museum Display* (Washington, D.C.: Smithsonian Institution

Press, 1991); Eilean Hooper-Greenhill, *Museums and the Shaping of Knowledge* (London: Routledge, 1992); Irit Rogoff, *Museum Culture: Histories, Discourses, Spectacles* (London: Routledge, 1994); Tony Bennett, *The Birth of the Museum: History, Theory, Politics* (London: Routledge, 1995).

5. A French expression connoting inherent—and in some cases inherited— knowledge. See Pierre Bourdieu's sociological studies of the connection between training in the appreciation of the arts and access to elite social status, most notably, *Distinction: A Social Critique of the Judgement of Taste* (Cambridge, Mass.: MIT Press, 1984).

6. Howard Gardner, *Frames of Mind: The Theory of Multiple Intelligences* (New York: Basic Books, 1983).

7. George E. Hein, Learning in the Museum (London: Routledge, 1998); John H. Falk and Lynn D. Dierking, *Learning from Museums: Visitor Experiences and the Making of Meaning* (Walnut Creek, Calif.: AltaMira, 2000).

8. An example is Andrew J. Pekarik, Zahava D. Doering, and David A. Karns, "Exploring Satisfying Experiences in Museums," *Curator* 42, no. 2 (1999): 152–73. An archive of visitor studies journals is available at the Visitor Studies Association website: www.visitorstudiesarchives.org/index.php (accessed June 27, 2008).

9. Madeleine Grynsztejn, interview with the author, San Francisco, July 5, 2007. Ms. Grynsztejn has since been appointed Pritzker Director of the Museum of Contemporary Art, Chicago.

10. For more on this point, including statistics from one study on the relative efficacy of introductory wall texts compared to other interpretive resources, see Peter Samis, "Gaining Traction in the Vaseline: Visitor Response to a Multi-track Interpretation Design for Matthew Barney: Drawing Restraint" (paper presented at Museums and the Web, San Francisco, April 2007), at www.archimuse.com/mw2007/papers/samis/samis.html (accessed June 27, 2008).

11. The first multiple-voice museum tour in the United States was produced by Antenna Audio in 1986 for the exhibition Bronislava Nijinska, A Dancer's Legacy, curated by Nancy Van Norman Baer at the de Young Museum in San Francisco. "Nancy and I went all over the country (mostly New York), interviewing various dancers like Frederick Franklin and Nina Youskevich who danced for Diaghilev in the Ballets Russes and knew Nijinsky very well. It was a great tour. Still one of the best, full of music and wonderful Russian voices telling great stories of the era and all the great choreographers and artists." Chris Tellis, founder, Antenna Audio, personal communication, August 27, 2007.

12. Moreover, recent research has shown that visitors benefit from having multiple complementary interpretive resources at their disposal. Visitors to the 2006 Matthew Barney exhibition at SFMOMA used as many as six different resources, choosing among introductory wall text, audio tour (in three formats), brochure, and a learning lounge that offered an artist video, interactive kiosks, wall graphics, and books to consult. Statistics revealed that among those unfamiliar with the artist, appreciation of the exhibition rose dramatically in proportion to the number of interpretive resources used. See Samis, "Gaining Traction in the Vaseline," and the study it was based on, Randi Korn et al., *Matthew Barney: Drawing Restraint Interactive Educational Technologies and Interpretation Initiative Evaluation* (San Francisco: SFMOMA, 2006), at www.sfmoma.org/whoweare/research_projects/ barney/RKA_2006_SFMOMA_Barney_distribution.pdf (accessed August 10, 2007).

13. Randy Kennedy, "With Irreverence and an iPod: Recreating the Museum Tour," *New York Times*, May 28, 2005, A1, at www.nytimes.com/2005/05/28/arts/design/ 28podc.html?ex=1274932800&en=db1ced6873dcc4b6&ei=5090&partner=rssuserland &emc=rss%20 (accessed August 6, 2007).

14. The exact language on the Web page was, "If you would like to share your audio guide with MoMA, e-mail us at audio@moma.org. MoMA will review submissions but reserves the right not to post them." In fact, they ended up posting one visitor-submitted set of audio tracks created in response to the 2006 Dada exhibition. The MoMA instructions for creating your own audio program are at www.moma.org/ visit_moma/createyourown.html (accessed September 15, 2007).

15. Visitors showing their MP3 player with an Artcast loaded receive a two dollar discount at the admission booth. The SFMOMA Artcast site and archive is at www.sfmoma.org/artcasts. For more on museum podcasting in general and the development of SFMOMA's solution in particular, see Peter Samis and S. Pau, "'Artcasting' at SFMOMA: First-Year Lessons, Future Challenges for Museum Podcasters" (paper presented at Museums and the Web, Albuquerque, N.M., April 2007), at www.archimuse.com/mw2006/papers/samis/samis.html (accessed August 6, 2007).

16. Available at www.tate.org.uk/youngtate/podcast/artlookers and http://redstudio .moma.org/podcasts/2006/index.php, respectively (accessed August 27, 2007).

17. Jennifer Trant, "Exploring the Potential for Social Tagging and Folksonomy in Art Museums: Proof of Concept," *New Review of Hypermedia and Multimedia* [PDF preprint]. For more on steve, see www.steve.museum. Other related initiatives are taking place as of this writing at the Powerhouse Museum in Sydney, Australia, at

www.powerhousemuseum.com/collection/database; the Smithsonian Photography Initiative's "Enter the Frame" project, at http://photography.si.edu/; and the Philadelphia Museum of Art, at www.philamuseum.org/collections/socialTagging .html (accessed June 26, 2008).

18. Olafur Eliasson, interview with author, Berlin, June 18, 2007.

19. Thomas H. Davenport and John C. Beck, *The Attention Economy: Understanding the New Currency of Business* (Cambridge, Mass.: Harvard Business School Press, 2001); and Peter J. Denning, "The Profession of IT: Infoglut," *Communications of the ACM* 49, no. 7 (July 2006): 16–19, at http://cs.gmu.edu/cne/pjd/PUBS/CACMcols/ cacmJul06.pdf (accessed June 26, 2008).

20. Chris Anderson, SJMA manager of interactive technology, personal communication, August 27, 2007.

21. The statement was made at the Open Circuits conference at the Museum of Modern Art in New York. John Baldessari, "TV (1) Is Like a Pencil and (2) Won't Bite Your Leg," in *The New Television: A Public/Private Art*, ed. Douglas Davis and Allison Simmons (Cambridge, Mass.: MIT Press, 1977), 110. Cited online by Cynthia Chris, "Video Art: Stayin' Alive," *Afterimage* (March 2000), at http://findarticles.com/ p/articles/mi_m2479/is_5_27/ai_61535391/pg_1 (accessed August 6, 2007).

22. For more examples of such "storytelling kiosks," see Brad Larson's blog entries at http://weblog.bradlarson.com/storytelling_kiosk/index.html (accessed August 28, 2007). In a recent phone conversation, Larson pointed out that, because children's and discovery museums have traditionally been collection free, they have always focused first and foremost on visitor experience, and have built up an extensive body of research. Art museums may have a lot to learn from them.

23. Paul Rudman et al., "Cross-Context Learning," in the present volume, chapter 9. See also Kevin Walker, "Visitor-Constructed Personalized Learning Trails," in *Museums and the Web 2007: Proceedings*, ed. Jennifer Trant and David Bearman (Toronto: Archives and Museum Informatics, 2007), at www.archimuse.com/ mw2007/papers/walker/walker.html (accessed August 28, 2007); and his chapter 7 in the present volume.

24. Vincent Puig and Xavier Sirven, "*Lignes de Temps*: Involving Cinema Exhibition Visitors in Mobile and On-line Film Annotation" (paper presented at Museums and the Web 2007, San Francisco, April 2007), at www.archimuse.com/mw2007/papers/ puig/puig.html (accessed November 5, 2007).

25. The seminal article on connectivism is George Siemens, "Connectivism: A Learning Theory for the Digital Age," December 4, 2004, at www.elearnspace.org/Articles/connectivism.htm (see also the connectivism site, blog, and wiki at www.connectivism.ca/ [accessed August 28, 2007]); David Weinberger, *Everything Is Miscellaneous: The Power of the New Digital Disorder* (New York: Times Books/Henry Holt, 2007). See also the website, at www.everythingismiscellaneous.com/ (accessed August 28, 2007).

26. Clearly, the figures are cumulative and not mutually exclusive: the top 13 percent, the "alpha bloggers," do all of these activities, while levels of engagement fall off after that. See www.powerhousemuseum.com/dmsblog/index.php/2007/04/23/more-on-levels-of-participation-forresters-social-technographics (accessed August 28, 2007). Similar data was reported in a report by the Pew Center for the Study of the Internet and American Life.

27. Samis, "Gaining Traction in the Vaseline"; and Korn et al., *Matthew Barney.*

28. Bonnie Pitman, "Serving Visitors with Choices: How Far Can Art Museums Stretch?" (presentation on a panel at American Association of Museums Annual Meeting, Chicago, May 2007).

29. George Siemens, "Connectivism: Museums as Learning Ecologies" (presentation to the Canadian Heritage Information Network's Roundtable on e-Learning, March 2006), at www.elearnspace.org/media/CHIN/player.html (accessed August 30, 2007).

2

Enhancing Visitor Interaction and Learning with Mobile Technologies

JOHN H. FALK AND LYNN D. DIERKING

Ask a visitor about his or her expectations for a museum visit and you find that whether visiting as part of a family, an all-adult group, or alone, she or he agrees on a few aspects: (1) the best museum is the one that presents a variety of interesting material and experiences that appeal to different age groups, educational levels, personal interests, and technical levels; (2) whether visiting as a group or alone, visitors expect to be mentally, and perhaps physically, engaged in some way by what they see and do; in other words, they expect to be able to personally connect in some way with the objects, ideas, and experiences provided; and (3) people visiting in groups, either families with children or all-adult groups, expect to enjoy a shared experience, as the members of the group with their varying interests and backgrounds collaborate and converse together.

In addition, there are some implicit assumptions. Paramount is the understanding that a museum has "the real stuff" (or reproductions at least), or is about the real stuff; consequently, visitors believe that there is an inherent sense of integrity to the objects, ideas, and experiences presented within the museum.[1] Another implicit assumption is that museums are, first and foremost, free-choice learning environments, that is, public institutions for personal learning,[2] as the visitor expectations shared above confirm so well. Thus, it is not surprising to us that, as Loïc Tallon highlights in his introduction, handheld technologies were among the first technologies to be fully embraced within museums and remain the most successful museum technology. This form of media represents an important and powerful way that museums can offer choice and individualized learning options to visitors, in the same way that a well-trained facilitator or well-designed hands-on exhibition offers such personalization. Digital media, well designed and wisely used, are important tools that can enhance visitor interaction and learning in museums, ensuring that these environments inspire and provoke curiosity and further understanding among visitors with varying backgrounds, interests, and knowledge levels.

Although at a gut level the field embraces these ideas, many questions remain. Why and in what ways do digital media tools such as multimedia tours, podcasts, interactive tours, intelligent tours, and remote Internet-enabled tours improve the learning potential of a museum visit? What inherent weaknesses are embodied in personal or handheld technologies? Is there evidence for how such technologies affect visitor learning when compared to more traditional media? In this chapter, we discuss a framework for thinking about learning in and from museums using mobile technologies.

Our thesis is that the key to understanding why well-designed digital media tools improve the learning potential of museum visits resides in understanding the nature of how visitors make meaning in and from museums—the mechanism by which people construct an understanding and appreciation of their world. We frame our discussion around the Contextual Model of Learning, a framework we developed some years ago but that we, and others, have continued to refine and elaborate upon.[3] We believe this framework broadly encapsulates the complexity of museum-based learning.

CONTEXTUAL MODEL OF LEARNING

The Contextual Model of Learning is not a model in the truest sense; it does not purport to make predictions other than that learning, or as we prefer to say these days, meaning making, is always a complex phenomenon situated within a series of contexts. The view of learning embodied in this framework is that individuals can be conceptualized as being involved in a continuous, contextually driven effort to make meaning in order to survive and prosper within the world, an effort that is best viewed as a never-ending dialogue between the individual and his or her physical and sociocultural environment. The Contextual Model of Learning portrays this contextually driven dialogue as both process and product of the interactions over time between three contexts: the personal, sociocultural, and physical contexts. None of these three contexts are ever stable or constant; all are changing across the lifetime of the individual. The Contextual Model of Learning draws from constructivist, cognitive, as well as sociocultural theories. The key feature of this framework is the emphasis on context.[4]

The personal context represents the sum total of personal and genetic history that an individual carries with him or her into a meaning-making situation. Building upon constructivist theories of learning, the influences of prior knowledge and experience on museum meaning making have been widely described and documented;[5] so, too, has the role of prior interest.[6] The nature of a visitor's motivations, or "identity-related agenda," for visiting a museum has also been shown to significantly influence the visitor's meaning-making outcomes.[7] More recently, it has been appreciated that the degree of choice and control also affects visitor meaning making.[8] Thus, from the personal context perspective, one should expect meaning making to reflect the realities of an individual's motivations and expectations, which in the case of museums typically involves a brief, usually leisure-oriented, culturally defined experience. One should also expect meaning making to be highly personal and strongly influenced by an individual's past knowledge, interests, and beliefs. And finally, one should expect meaning making to be influenced by an individual's desire to both select and control his or her own experiences in order to fulfill identity-related needs.

The sociocultural context recognizes that humans are innately social creatures, products of their culture and social relationships.[9] Hence, one should expect museum meaning making to be always socioculturally situated. Sociocultural factors hypothesized to affect meaning making include such large-scale influences as the cultural value placed upon free-choice learning itself,[10] as well as the cultural context of the museum within society.[11] Although these factors clearly influence the meaning made in these settings, empirical evidence for these impacts is difficult to find. However, considerable research exists showing that visitors to museums are strongly influenced by the interactions and collaborations they have with individuals within their own social group.[12] Research has also shown that the quality of interactions with others outside the visitor's own social group—for example, museum explainers, guides, demonstrators, performers, or even other visitor groups—can make a profound difference in visitor meaning making.[13]

Finally, meaning making always occurs within a physical setting; in fact, it is always a dialogue with the physical environment. Thus the third context is the physical context. One expects—in fact hopes given the effort museums invest in design and architecture—that visitors to museums will notice and respond to the physical context of the museum itself: both the large-scale properties of space, lighting, and climate, as well as microscale aspects such as the exhibitions and specific objects contained within them. Utilizing the Contextual Model of Learning framework, personal handheld devices represent a part of the physical context of the visitor.

Since museums are typically free-choice settings, the experience is generally voluntary, nonsequential, and highly responsive to what the setting affords.[14] As such, visitor meaning making has been shown to be strongly influenced by how successfully visitors are able to orient themselves within the space, since being able to confidently navigate within a complex three-dimensional environment turns out to be highly correlated with what and how much an individual learns.[15] Similarly, intellectual navigation has been shown to affect visitor meaning making from museums.[16] Research has also shown that a myriad of architectural design factors such as lighting, crowding, color, sound, and space subtly influence visitor meaning making.[17] And not least, considerable research has focused on the technology, exhibitions, objects, and labels themselves, since they are designed to be the primary focus of most museums. Not surprisingly then, ample evidence exists that instructional design features influence meaning making, in particular the sequencing, positioning, and content of media, exhibitions, and labels,[18] how many educational elements a visitor attends to (and for how long),[19] and the nature of the museum objects and displays themselves.[20]

Finally, less well documented but theoretically compelling is the expectation that meaning making from museums will not only be influenced by the reinforcement and enrichment of visitors' previously known intellectual constructs but also equally depend upon what happens subsequently in the learner's environment, since meaning making is not an instantaneous phenomenon but rather a cumulative process of acquisition and consolidation.[21] Thus, experiences occurring after the visit frequently play an important role in determining, in the long term, what meaning is actually "made" in the museum. Demonstrating the importance of this aspect of the museum

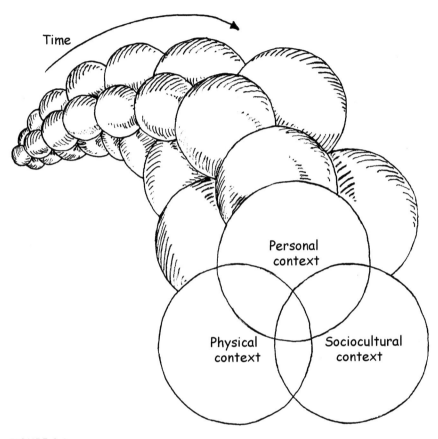

FIGURE 2.1
Contextual Model of Learning.

experience, the American Association of Museums' MUSE Award now has an "Extended Experience" category.

These three contexts provide a large-scale framework with which to organize the complexity of meaning making. Within the framework are myriad factors, probably numbering in the hundreds, if not thousands. Some of these factors are apparent and have been summarized above and in previous publications;[22] others are either not apparent or not currently perceived by the field to be important. After considering the findings from hundreds of research studies including the ones cited above, twelve key factors—or more accurately suites of factors—emerge as influential for museum meaning-making experiences. These twelve factors are as follows:

Personal Context
- Visit motivation and expectations
- Prior knowledge and experience
- Prior interests
- Choice and control

Sociocultural Context
- Cultural background
- Within group social mediation
- Mediation by others outside the immediate social group

Physical Context
- Advance organizers
- Orientation to physical space
- Architecture and macroscale environmental factors
- Design of exhibitions, programs, and technology
- Subsequent reinforcing events and experiences outside the museum

PIECES AND WHOLES

Research has shown that these twelve factors contribute to the quality of a museum experience, though the relative importance of any one factor may vary between particular visitors and venues (e.g., art museums vs. natural history museums, families vs. all-adult groups, and so on). A recent study found that all of these factors emerged as having a significant, though often small, correlation with change in meaning making (as measured across nearly a dozen different dimensions).[23] As the Contextual Model of Learning proposes, the only way to fully make sense of the museum experience is to consider all variables simultaneously. In other words, all of these factors were important; however, no single factor adequately accounted for visitor meaning-making outcomes across all of the nearly two hundred visitors we studied. Even the "best" of these factors explained only about 9 percent of the variance in meaning making. The data support the idea that collections of factors, rather than individual factors, provide a reasonable explanation for the nature of visitor changes in understanding, attitudes, appreciation, and beliefs.

Perhaps most importantly, different suites of factors were found to significantly affect different groups of visitors.

The key to teasing out these trends was in being able to meaningfully segment visitor groups. Traditional demographic categories like age, race/ethnicity, and even social group and educational attainment proved of limited usefulness in this respect. By contrast, grouping visitors as a function of their prior knowledge, motivation for visiting the museum, or prior interest in the topic of the museum was useful. These groupings allow us to begin to tease out how different factors could be important for some groups and not others. For example, for visitors with the most limited entering understanding of the content of the exhibition, exhibit quality emerged as a highly significant contributor to meaning making; orientation to the exhibition space had no significant effect on this group.

These findings suggest that individuals with limited entering knowledge might be a key group for use of content-focused handheld devices. However, exhibit quality was not an important factor for visitors with a reasonable amount of prior understanding of the topic; for this group, orientation to the exhibition was most significant to their meaning making. Again, extending this finding to digital technologies, this group might particularly benefit from opportunities to plan their trip in advance by downloading orientation information onto a user-owned handheld technology—cell phone or MP3 player—from the museum website. Clearly, though, applying these findings implies the ability to steer visitors to the cues and supports that best meet their needs, which in turn implies being able to segment visitors according to these criteria. Currently, that capability does not exist, though a number of institutions are experimenting with strategies for doing this, and other chapters in this book explore some of these.

Overall, the results of this major investigation demonstrate that meaning making is influenced by a wide range of factors. That said, even when all of the variables are considered, they do not provide a complete picture; at best, the researchers could only explain about half of the variance—a fraction of why different individuals left making the meaning they did. However, one thing seems clear: the personal context of the visitor—his or her prior experiences, knowledge, motivations for visiting in the first place, and interests—is more important than any particular exhibition technique or innovation in predicting learning outcomes. This does not mean that physical context variables such as what people see and do, including whether or not they utilize digital technologies, do not influence visitor learning at all—quite the contrary. However, it does mean that visitor learning will only be enhanced by such technologies to the degree to which visitors can use them as tools to personally customize their visit in ways that build on and optimize their prior experiences, visit motivations, and interests.

EXTENDING THE SCOPE AND SCALE OF THE MUSEUM EXPERIENCE

Implicit in the Contextual Model of Learning is that visitor learning is not only an in-museum experience. For despite being feasible, practical, and outwardly reasonable to conceptualize the role and impacts of digital media on museum experiences as being delimited by the physical scope and time scale of the actual museum visit, a growing body of research suggests that museum learning is not confined to the in-museum experience. Understanding meaning making requires taking a much longer view than just the couple of hours an individual spends within a museum.[24] In this view, meaning making is a whole, not a part—a whole that can only be understood by trying to situate any given meaning-making experience within the larger framework of a person's entire life. In other words, to document a meaning-making situation requires panning the camera back in time and space in order to see how this particular situation fits within the larger picture of a person's life.

Appreciation of the importance of scope and scale amongst investigators of museum meaning making has been slow in coming, and most efforts to accommodate these ideas are still quite crude. In addition, there is another issue. Although some in the field still assume that meaning making from museums means the addition of "new" understanding on the part of visitors, there is a growing awareness that much of the meaning making afforded by museums is in the area of consolidation and reinforcement of previous understandings, and individuals are much more likely to describe the outcomes of their museum experiences as strengthening rather than changing their existing knowledge structures.[25] For example, in random telephone interviews of Los Angeles residents, 95 percent of the individuals reporting previously visiting a specific California museum believed that their museum experience had strengthened or extended their knowledge of some topic presented by the museum, while, in a separate question, only 66 percent felt that their museum experience had changed their understanding, attitudes, or behaviors.[26]

Over the past decade, a growing number of museum investigators have begun to incorporate longer time frames into their research. Recent longitudinal
studies have shown that the meaning making that results from a museum experience does change over time and not always by declining.[27] What these
studies reinforce is that long-term outcomes are often not predictable from
short-term outcomes. Thus, in order to understand visitor's museum meaning making requires understanding visitors across three time periods: (1) visitor pre-museum history—visitors' prior knowledge, interest, experience,
expectations, and identity-related motivations; (2) in-museum experiences—
the actual experiences visitors engage in with specific exhibitions, their social
interactions both within and outside their own social group, as well as characteristics of the physical setting such as crowdedness and the presence or absence of advance organizers; and (3) visitor post-museum history—the types
of reinforcing experiences visitors have such as post-visit conversations, reading, television watching, and so forth. Recent research has begun to indicate
that, at least for some visitors, all of these time frames are inextricably inter-
twined.[28] In other words, what visitors "bring with them" to the museum directly influences what they do while there, but equally important, these
entering conditions also create a filter through which individuals frame their
memories of the museum experience months and years later.

THE CONTEXTUAL MODEL AND DIGITAL MEDIA

This research reinforces how incredibly complex the museum experience is
and offers insights into why and in what ways digital media tools have the potential to enhance the meaning made of and from these experiences. Clearly,
physical context variables such as the design of media tools and the organization and navigation of the content presented or afforded by these tools are important, but of equal—if not greater—importance are a visitor's personal and
sociocultural contexts. And since visitors do not make meaning from museums solely within the four walls of the institution, effective digital media experiences require situating the experience within the broader context of the
lives, the community, and the society in which visitors live and interact.

This does not mean that digital technologies do not influence visitor learning; there is certainly preliminary evidence that they do. In particular, we feel
that such technologies, when designed well, can have the potential to positively impact visitor meaning making, by (1) enabling visitors to customize

their experiences to meet their personal needs and interests; (2) extending the experience beyond the temporal and physical boundaries of the museum visit; and (3) layering multisensory elements within the experience, thereby enriching the quality of the physical context.

However, as is highlighted in Tallon's introduction, there has been only a smattering of research or evaluation conducted on the educational effectiveness of digital technologies—in particular, the real value added by these technologies as compared to some other educational intervention. There are a few studies focused on learning in art museums,[29] science centers,[30] and aquaria.[31] Little empirical research has been conducted, however, that directly compares the educational outcomes resulting from the use of digital technologies with the educational outcomes achieved using more traditional media; in other words, whether digital media are superior to other forms of intervention. It is not sufficient that new technologies enhance the visitor experience; it needs to be demonstrated that these new technologies enhance the visitor experience better than competing technologies and in ways that are cost-effective. It is not until the field has a strong research base that it will truly be able to both optimize the power of these digital media tools and substantiate their value. It seems fair to say that a full understanding of how digital technologies support museum-based meaning making lies more in the future than in the present. However, that future is not likely too distant.

The museum experience is incredibly complex. Clearly, physical context variables such as the design of digital media tools are important, but equally if not more important are a visitor's personal and sociocultural contexts. Digital media experiences have the potential to effectively situate the visitor's museum experience within the broader context of an individual's life, community, and society; they also have the potential to allow significant customization of experience and to extend visitor experiences beyond the temporal and physical boundaries of the institution. However, this potential to enhance the quality of the visitor's experiences, including efforts designed to advance the museum's educational mission, will only be achieved to the degree technologies actually fulfill the personal and social needs of the visitor. Specifically, technologies must build on and optimize visitor's prior experiences and knowledge, connect to their social group, and directly support visitor's motivations for visiting and their interests before, during, and after the experience.

NOTES

1. John H. Falk and Lynn D. Dierking, *The Museum Experience* (Washington, D.C.: Whalesback Books, 1992).

2. John H. Falk and Lynn D. Dierking, eds. *Public Institutions for Personal Learning* (Washington, D.C.: American Association of Museums, 1995).

3. John H. Falk and M. Storksdieck, "Using the Contextual Model of Learning to Understand Visitor Learning from a Science Center Exhibition," *Science Education* 89 (2005): 744–78.

4. This framework for thinking about meaning making has also been emphasized by others (e.g., S. J. Ceci, *On Intelligence: A Bioecological Treatise on Intellectual Development* (Boston: Harvard University Press, 1996); S. J. Ceci and U. Bronfenbrenner, "Don't Forget to Take the Cupcakes out of the Oven: Strategic Time Monitoring, Prospective Memory, and Context," *Child Development* 56 (1985): 175–90; R. J. Sternberg and R. K. Wagner, eds., *Practical Intelligence: Nature and Origins of Competence in the Everyday World* (Cambridge, UK: Cambridge University Press, 1996).

5. Lynn D. Dierking and W. Pollock, *Questioning Assumptions: An Introduction to Front-End Studies* (Washington, D.C.: Association of Science Technology Centers, 1998); John H. Falk and L. Adelman, "Investigating the Impact of Prior Knowledge, Experience and Interest on Aquarium Visitor Learning," *Journal of Research in Science Teaching* 40, no. 2 (2003): 163–76; Rochel Gelman, Christine M. Massey, and Mary McManus, "Characterizing Supporting Environments for Cognitive Development: Lessons from Children in a Museum," in *Perspectives on Socially Shared Cognition*, ed. Lauren B. Resnick, John M. Levine, and Stephanie D. Teasley (Washington, D.C.: American Psychological Association, 1991), 226–56; George E. Hein, *Learning in the Museum* (London: Routledge, 1998); Jeremy Roschelle, "Learning in Interactive Environments: Prior Knowledge and New Experience," in *Public Institutions for Personal Learning*, ed. John H. Falk and Lynn D. Dierking (Washington, D.C.: American Association of Museums, 1995), 37–51; L. Silverman, "Making Meaning Together," *Journal of Museum Education* 18, no. 3 (1993): 7–11.

6. L. M. Adelman, L. D. Dierking, K. Haley Goldman, D. Coulson, J. H. Falk, and M. Adams, *Baseline Impact Study: Disney's Animal Kingdom Conservation Station*, technical report (Annapolis, Md.: Institute for Learning Innovation, 2001); Mihaly Csikszentmihalyi and K. Hermanson, "Intrinsic Motivation in Museums: Why Does One Want to Learn?" in *Public Institutions for Personal Learning*, ed. John H. Falk and Lynn D. Dierking (Washington, D.C.: American Association of Museums, 1995), 67–78; Falk and Adelman, "Investigating the Impact."

7. John H. Falk, "Field Trips: A Look at Environmental Effects on Learning," *Journal of Biological Education* 17, no. 2 (1983): 134–42; John H. Falk, "The Impact of Visit Motivation on Learning: Using Identity as a Construct to Understand the Visitor Experience," *Curator* 49, no. 2 (2006): 151–66; John H. Falk, Theanno Moussouri, and D. Coulson, "The Effect of Visitors' Agendas on Museum Learning," *Curator* 41, no. 2 (1998): 106–20; John H. Falk, J. E. Heimlich, and K. Bronnenkant, "The Identity-Related Motivations of Adult Zoo and Aquarium Visitors," *Curator*, forthcoming; Nelson H. Graburn, "The Museum and the Visitor Experience," in *The Visitor and the Museum* (Seattle: 72nd Annual Conference of the American Association of Museums, 1977), 5–32; M. Hood, "Staying Away: Why People Choose Not to Visit Museums," *Museum News* 61, no. 4 (1983): 50–57; J. Packer, "Learning for Fun: The Unique Contribution of Educational Leisure Experiences," *Curator* 49, no. 3 (2006): 329–44; J. Packer and R. Ballantyne, "Motivational Factors and the Visitor Experience: A Comparison of Three Sites," *Curator* 45 (2002): 183–98.

8. Jeanette Griffin, "School-Museum Integrated Learning Experiences in Science: A Learning Journey" (unpublished PhD diss., University of Technology, Sydney, 1998); R. B. Lebeau, P. Gyamfi, K. Wizevich, and E. H. Koster, "Supporting and Documenting Choice in Free-Choice Science Learning Environments," in *Free-Choice Science Education: How We Learn outside of School*, ed. John H. Falk (New York: Teachers College Press, 2001), 133–48.

9. John Ogbu, "The Influence of Culture on Learning and Behavior," in *Public Institutions for Personal Learning: Establishing a Research Agenda*, ed. John H. Falk and Lynn D. Dierking (Washington, D.C.: American Association of Museums, 1995), 79–96; James V. Wertsch, *Vygotsky and the Social Formation of the Mind* (Cambridge, Mass.: Harvard University Press, 1985).

10. Ogbu, "The Influence of Culture"; Wertsch, *Vygotsky and the Social Formation*.

11. M. Bal, *Double Exposures* (London: Routledge, 1996); Tony Bennett, *The Birth of the Museum: History, Theory, Politics* (London: Routledge, 1995); Eilean Hooper-Greenhill, *Museums and the Shaping of Knowledge* (London: Routledge, 1992).

12. Minda Borun, Matthew Chambers, J. Dritsas, and J. Johnson, "Enhancing Family Learning through Exhibits," *Curator* 40, no. 4 (1997): 279–95; Kevin Crowley and Maureen Callanan, "Describing and Supporting Collaborative Scientific Thinking in Parent-Child Interactions," *Journal of Museum Education* 23, no. 1 (1998): 12–17; Kirsten M. Ellenbogen, "Museums in Family Life: An Ethnographic Case Study," in *Learning Conversations in Museums*, ed. Gaea Leinhardt, Kevin Crowley, and Karen Knutson (Mahwah, N.J.: Erlbaum, 2002), 81–101; L. Schaubel, D. Banks, G. D.

Coates, L. M. W. Martin, and P. Sterling, "Outside the Classroom Walls: Learning in Informal Environments," in *Innovations in Learning*, ed. L. Schauble and R. Glaser (Mahwah, N.J.: Erlbaum, 1996), 5–24.

13. T. Astor-Jack, K. K. Whaley, Lynn D. Dierking, D. Perry, and C. Garibay, "Understanding the Complexities of Socially Mediated Learning," in *In Principle, In Practice: Museums as Learning Institutions*, ed. John H. Falk, Lynn D. Dierking, and S. Foutz (Linthicum, Md.: AltaMira, 2007), 217–28; Crowley and Callanan, "Describing and Supporting"; J. J. Koran Jr., M. L. Koran, Lynn D. Dierking, and J. Foster, "Using Modeling to Direct Attention in a Natural History Museum," *Curator* 31, no. 1 (1988): 36–42; I. Wolins, N. Jensen, and R. Ulzheimer, "Children's Memories of Museum Field Trips: A Qualitative Study," *Journal of Museum Education* 17, no. 2 (1992): 17–27.

14. John H. Falk and Lynn D. Dierking, *Learning from Museums: Visitor Experiences and the Making of Meaning* (Walnut Creek, Calif.: AltaMira, 2000).

15. For example, G. Evans, "Learning and the Physical Environment," in *Public Institutions for Personal Learning*, ed. John H. Falk and Lynn D. Dierking (Washington, D.C.: American Association of Museums, 1995), 119–26; D. G. Hayward and M. Brydon-Miller, "Spatial and Conceptual Aspects of Orientation: Visitor Experiences at an Outdoor History Museum," *Journal of Environmental Systems* 13, no. 4 (1984): 317–32; C. A. Kubota and R. G. Olstad, "Effects of Novelty-Reducing Preparation on Exploratory Behavior and Cognitive Learning in a Science Museum Setting," *Journal of Research in Science Teaching* 28, no. 3 (1991): 225–34.

16. This is supported by quality advance organizers. See David Anderson and Keith B. Lucas, "The Effectiveness of Orienting Students to the Physical Features of a Science Museum Prior to Visitation," *Journal of Research in Science Teaching* 27, no. 4 (1997): 485–95; John H. Falk, "Testing a Museum Exhibition Design Assumption: Effect of Explicit Labeling of Exhibit Clusters on Visitor Concept Development," *Science Education* 81, no. 6 (1997): 679–88.

17. J. Coe, "Design and Perception: Making the Zoo Experience Real," *Zoo Biology* 4 (1985): 197–208; Evans, "Learning and the Physical Environment; A. Hedges, "Human-Factor Considerations in the Design of Museums to Optimize Their Impact on Learning," in *Public Institutions for Personal Learning*, ed. John H. Falk and Lynn D. Dierking (Washington, D.C.: American Association of Museums, 1995), 105–18; J. L. Ogden, D. G. Lindburg, and T. L. Maple, "The Effects of Ecologically Relevant Sounds on Zoo Visitors," *Curator* 36, no. 2 (1993): 147–56.

18. Lynn D. Dierking and John H. Falk, "Audience and Accessibility," *The Virtual and the Real: Uses of Multimedia in Museums*, ed. S. Thomas and A. Mintz (Washington, D.C.: Technical Information Services, American Association of Museums, 1998), 57–72; Steve Bitgood and D. Patterson, "Principles of Exhibit Design," *Visitor Behavior* 2, no. 1 (1995): 4–6; John H. Falk, "Assessing the Impact of Exhibit Arrangement on Visitor Behavior and Learning," *Curator* 36, no. 2 (1993): 1–15; Beverly Serrell, *Exhibit Labels: An Interpretive Approach* (Walnut Creek, Calif.: AltaMira, 1996).

19. Steve Bitgood, Beverly Serrell, and D. Thompson, "The Impact of Informal Education on Visitors to Museums," in *Informal Science Learning: What Research Says about Television, Science Museums, and Community-Based Projects*, ed. V. Crane (Dedham, Mass.: Research Communications, 1994), 61–106; Beverly Serrell, *Paying Attention: Visitors and Museum Exhibitions* (Washington, D.C.: American Association of Museums, 1998).

20. S. Paris, ed., *Perspectives on Object-Centered Learning in Museums* (Mahwah, N.J.: Erlbaum, 2002).

21. David Anderson, "Understanding the Impact of Post-Visit Activities on Students' Knowledge Construction of Electricity and Magnetism as a Result of a Visit to an Interactive Science Centre" (unpublished PhD diss., Queensland University of Technology, Brisbane, Australia, 1999); J. D. Bransford, A. L. Brown, and R. Cocking, eds., *How People Learn: Brain, Mind, Experience, and School* (Washington, D.C.: National Research Council, 1999); M. I. Medved, "Remembering Exhibits at Museums of Art, Science and Sport" (unpublished PhD diss., University of Toronto, 1998).

22. Falk and Dierking, *Learning from Museums*; Falk and Storksdieck, "Using the Contextual Model."

23. Falk and Storksdieck, "Using the Contextual Model."

24. Falk and Storksdieck, "Using the Contextual Model."

25. Anderson, "Understanding the Impact"; Kirsten M. Ellenbogen, "From Dioramas to the Dinner Table: An Ethnographic Case Study of the Role of Science Museums in Family Life" (unpublished PhD diss., Vanderbilt University, Nashville, 2003); Falk and Dierking, *Learning from Museums*.

26. John H. Falk, P. Brooks, and R. Amin, "Investigating the Long-Term Impact of a Science Center on Its Community: The California Science Center L.A.S.E.R. Project,"

in *Free-Choice Science Education: How We Learn Science outside of School*, ed. John H. Falk (New York: Teachers College Press, 2001), 115–32.

27. Anderson, "Understanding the Impact"; L. M. Adelman, J. H. Falk, and S. James, "Assessing the National Aquarium in Baltimore's Impact on Visitor's Conservation Knowledge, Attitudes and Behaviors," *Curator* 43, no. 1 (2000): 33–62; S. Bielick and D. Karns, *Still Thinking about Thinking: A 1997 Telephone Follow-up Study of Visitors to the Think Tank Exhibition at the National Zoological Park* (Washington, D.C.: Institutional Studies Office, Smithsonian Institution, 1998); Ellenbogen, "Museums in Family Life"; John H. Falk et al., "Interactives and Visitor Learning," *Curator* 47 (2004): 171–98; J. Luke, M. Cohen Jones, Lynn D. Dierking, M. Adams, and John H. Falk, *The Impact of Museum Programs on Youth Development and Family Learning: The Children's Museum, Indianapolis Family Learning Initiative—Technical Report* (Annapolis, Md.: Institute for Learning Innovation, 2002); Medved, "Remembering Exhibits."

28. Falk, Heimlich, and Bronnenkant, "The Identity-Related Motivations"; John H. Falk and M. Storksdieck, "Learning Science from a Leisure Experience: Understanding the Long-Term Meaning Making of Science Center Visitors," *Journal of Research in Science Teaching*, forthcoming.

29. Nancy Proctor, Jane Burton, and Chris Tellis, "The State of the Art in Museum Handhelds in 2003," in Museums and the Web 2003: Selected Papers from an International Conference, Charlotte, N.C., March 2003, at www.archimuse.com/mw2003/papers/proctor.html (accessed April 2007); Nancy Proctor and Jane Burton, "Tate Modern Multimedia Tour Pilots 2002–2003 (unpublished evaluation report from Tate Modern, London, 2003).

30. Sherry Hsi, "I-Guides in Progress: Two Prototype Applications for Museum Educators and Visitors using Wireless Technologies to Support Informal Science Learning," in *Proceedings of the 2nd IEEE International Workshop on Wireless and Mobile Technologies in Education* (JungLi, Taiwan, 2004), 187–92.

31. F. Bellotti, R. Berta, A. de Gloria, and M. Margarone, "User Testing a Hypermedia Tour Guide," *Pervasive Computing* 1, no. 2 (2002): 33–41.

3

Designing Mobile Digital Experiences

BEN GAMMON AND ALEXANDRA BURCH

Mobile digital technology offers museums, zoos, aquaria, botanical gardens, and art galleries many valuable and unique opportunities to increase access to their collections and to enhance their visitors' learning and enjoyment. Yet, realizing these opportunities is not straightforward, and there are many potential pitfalls along the way. Key to the success of mobile digital technology in a museum is a detailed understanding of the visitors' needs, wants, expectations, and behaviors. Research into visitors' reactions to and use of mobile digital technology—supplemented by similar research into fixed kiosk-based computer exhibits—is beginning to shed light upon these issues and to provide useful guidance as to how to develop genuinely effective mobile digital interpretation for museums.

WHAT CAN MOBILE DIGITAL TECHNOLOGY OFFER MUSEUM VISITORS?

Digital technology—in the form of static kiosk exhibits[1] or as handheld mobile devices[2]—offers museums many opportunities that traditional print-based or live interpretation cannot provide. It offers visitors access to vast resources of information from which they can search and select items of interest, providing museums with the opportunity to cater to people with different learning styles seeking very different types of information.[3] For example, the Exploratorium in San Francisco developed a series of video labels that presented different types of narrative or inquiry-based approaches to enhance their interactive exhibits.[4] The Victoria and Albert Museum in London (V&A) has used computer-based labels to provide greatly enriched interpretation for their collections of fine art objects.[5] Digital technology can provide these rich resources of information through a wide range of media—animations, video, static images, sound, and text—and, if needed, in many different languages, including sign languages.[6]

The potential of digital technology goes far beyond providing a high performance label. It can immensely enrich visitors' enjoyment and learning in ways that would be extremely difficult if not impossible to provide through other media. It can provide richly authentic learning experiences—activities and resources that are much closer to those found in the real world, and which cover topics more closely aligned to students' interests than those that can be delivered via traditional educational techniques.[7]

We propose that the advantages of digital technology center on its ability to connect users with other learners, to provide opportunities to explore and construct models of real-world systems, and to represent data in many different forms.

There are many innovative examples of such applications in museum settings. For example, digital video cameras have been used to record visitors' interactions with a miniature tornado. Visitors could then replay the video clips at normal speed, in slow motion, and in reverse, and this was found to dramatically increase visitors' engagement with the interactive exhibit, by helping them to identify salient aspects of the experience they had previously missed, and by encouraging them to return to the exhibit to explore the phenomenon in greater depth.[8]

Digital technology also allows museum visitors to engage in gameplay and exploration of experiences that would be impossible to replicate in the real world because such experiences are too small, too large, too slow, too fast, or too expensive. For example, a computer interactive at the "Energy—Fuelling the Future" exhibition at the Science Museum in London allows visitors to play the role of a minister of energy for an imaginary country. With this exhibit visitors can determine the energy policy over a twenty-five-year period and discover the impact of their decisions upon the country's economy and environment, as well as their political careers.[9]

Digital technology provides opportunities for museum visitors to capture and take back to home or school reminders of their museum experience, such as digitized images of objects or artworks, the results of activities they have taken part in, or links to sources of information for further study.[10] Such "bookmarking and retrieval" functions can help to extend the experience of the museum far beyond the confines of the actual visit, thereby dramatically increasing its potential impact, as John Falk and Lynn Dierking suggest in the previous chapter. This is especially advantageous for school groups, for whom it has been shown that post-visit activities in class are vital to ensuring sustained learning from the museum experience but which teachers often struggle to deliver.[11] (Chapters 5, 7, and 8 discuss bookmarking further.)

While it would be possible to provide some of the aforementioned experiences using other media, the breadth and quality of what can be provided, the potential to quickly and easily update information, and the size and breadth of the audiences that can be reached is far greater using digital technology.

Mobile digital technology provides all of these opportunities plus others unique to devices such as personal digital assistants (PDAs) or mobile phones. Unlike fixed kiosk-based computers, mobile digital technology can provide personalized interpretation since, in effect, each visitor has his or her own portable computer terminal. Visitors can access the information or activities they want, when they need them.

Since a handheld device is small and lightweight, the visitor can move it to the position best suited to simultaneously view the object and its accompanying interpretation.

Mobile digital devices also provide the opportunity to build communities of learners through the use of wireless networking, allowing visitors to communicate at a distance with others in or out of the museum, to share information, ideas, and experiences, or even to engage in multiplayer games.[12]

Many innovative uses of mobile digital devices have already been tested in schools[13] and to a lesser extent in informal learning settings including museums.[14] However, research studies have shown that, while there are many opportunities, there are also many potential pitfalls, and the successful application of mobile digital technology to museum settings requires a detailed understanding of visitors' often unpredictable needs, wants, expectations, and behaviors. In particular, there is a need to address the following issues:

- How do visitors react to the use of digital technology in different types of museum settings?

- Does the technology distract attention away from the museum's objects, artworks, and interactive exhibits?
- Does digital technology disrupt important social interactions that promote learning?
- Can visitors make effective use of the technology?

HOW DO VISITORS RESPOND TO DIGITAL TECHNOLOGY?

Concern has been expressed that the use of digital technology in museums is unpopular with visitors.[15] However, evidence from studies in many different museums clearly indicates that such concerns are unfounded when the design of the hardware, software, and content has been based upon an understanding of visitors' needs, wants, expectations, and patterns of behavior.

Museums' increasing use of digital technology has found wide support by the visiting public of all ages. Strong support is found particularly among visitors to science centers and science museums, where digital exhibits have been shown to be immensely popular, with high attracting and holding power.[16] For example, visitor tracking in two exhibitions at the Science Museum of Minnesota revealed that the two key factors that positively correlated with the amount of time spent at an exhibit were its technological novelty and open-endedness—both factors in which the digital technology exhibits were particularly strong.[17] It has also been found that visitors to science centers and museums expect and demand that contemporary science topics be conveyed using contemporary digital technology.[18]

While the support for the use of digital technology among visitors to science centers and museums is perhaps unsurprising, it is interesting to note that equally strong support is found among both adults and children visiting art museums and galleries. The Victoria and Albert Museum's British Galleries are a particularly interesting case study.[19] These galleries, opened in 2001, display a vast collection of objects illustrating applied art and design from the sixteenth to the nineteenth centuries. In addition to thousands of objects and artworks, the British Galleries make extensive use of video and interactive computer exhibits as well as mechanical interactive exhibits.

Despite concerns expressed during their development, the new galleries have been found to attract a substantially larger audience than the traditional displays of objects with printed labels that they replaced. Among visitors to the new galleries, there was no evidence of hostility to the use of computer exhibits. In a survey of 290 adult visitors to the British Galleries it was found that 87 percent felt the presence of computer exhibits was appropriate and beneficial.[20] Even among those visitors who did not use any of the interactive interpretation, 76 percent felt that the use of computer exhibits was appropriate and did not detract from the atmosphere of the exhibition. Overall, 94 percent of the visitors agreed that the British Galleries had a tranquil atmosphere, and 93 percent agreed with the statement that the galleries are "just like a museum should be."

In a follow-up study, it was found that over 90 percent of the visitors used at least one of the interactive exhibits in the British Galleries (with the average being five exhibits per visitor) and that some of the most popular exhibits were the computer kiosks, especially with the adult audience.[21]

Similarly, positive responses have been found for fixed computer exhibits and handheld digital devices in art galleries such as the National Gallery[22] and Tate Modern,[23] both in London. As far as visitors are concerned, digital technology is an entirely appropriate and much valued interpretative tool for museums.

DOES DIGITAL TECHNOLOGY DETRACT ATTENTION?

Regardless of its popularity with visitors, concerns have also been expressed that digital technology detracts visitors' attention from real objects and artworks. While some studies indicate that digital technology can be a distraction,[24] there is considerable counterevidence that when it is properly designed, it can actually increase visitors' engagement with other exhibits.

Two case studies from the Science Museum, London, of the use of static kiosk-based computer labels illustrate that, when properly designed, digital technology actually increases visitors' engagement with real objects. In the recently redesigned Energy Hall—a gallery displaying stationary steam engines from the Industrial Revolution[25]—new computer labels were found to have been highly successful in engaging visitors' interest in the historic objects. These digital labels contain computer animations of how the engines worked, as well as images and text-based information illustrating their historical context. Observation and interviews with over three thousand visitors revealed that 77 percent of those who engaged with the objects also used the accompanying computer label.[26] Half of these visitors started by engaging with the object and then moved to the computer label, indicating that the digital technology was being used as complementary interpretation rather than merely displacing attention away from the object. Two-thirds of the visitors who used the computer labels were observed to also engage deeply with the related object, compared to just 1 percent of visitors who did not use the computer labels.

Similar results were found for kiosk-based computer labels in the Science Museum's "Nanotechnology—Small Science, Big Deal" exhibition.[27] Half of visitors who were observed to look at an object also looked at the accompanying computer label, and most of these visitors were observed to alternate between looking at the computer screen and looking at the related object. Visitors were often observed using the computer labels together and discussing the information provided about the objects.[28]

The success of these computer labels was due to the following: their position and orientation so that visitors could easily view the related object while standing at the terminal; the use of on-screen images to indicate which object the computer label referred to; the title, design, and content of the initial screens that clearly indicated the purpose of the computer labels (i.e., as information points, not interactive games); and the use of multimedia to illustrate aspects of the object's construction, purpose, and operation that could not be gained from simply looking at the object or a printed label.

Studies of visitors to the V&A's British Galleries confirm these findings.[29] Visitors were observed to use the kiosk-based computer labels to seek more information about objects and were found to have engaged more deeply with the objects as a result of using them. Visitors themselves felt that the computer labels helped them to appreciate and understand more about the objects. As with the Science Museum case studies, it was found that the location and orientation of the exhibits in the British Galleries were crucial to their success. It was vital for visitors to be able to easily view the objects referred to by the computer labels as they stood in front of the kiosk.

As with static kiosk-based computer labels, mobile digital technology has been shown to be successful in focusing visitors' attention and increasing their engagement with real objects and artworks. For example, a mobile digital guide to a historical house was developed.[30] Comparing the observed behaviors and conversations of visitors using the mobile digital guide to those using a traditional paper guidebook, it was found that the digital guide was actually more effective in drawing and holding visitors' attention to real objects and generated more discussions about the objects.[31] Visitors' responses to the mobile digital guide were very positive, with users feeling that it had increased the time they spent looking at the objects and the amount they had learned about them.

The evaluation of PDA tours of Tate Modern's permanent galleries and of a temporary exhibition of Frida Kahlo's work also showed very positive responses from visitors.[32] More than 70 percent of visitors said that they had spent longer in the galleries as a result of using the PDA tour, and more than 70 percent felt that using it had improved their visit to the galleries by enhancing their appreciation of the artworks. Visitors responded particularly positively to audio tracks of artists talking about their work, videos showing how the artworks were produced, and digitized images that allowed them to examine parts of the artwork in greater detail.

Even in the radically different environment of a zoo, a prototype digital guided tour (delivered via mobile phone) was found to be popular with visitors and increased their engagement with the animal displays—especially for the enclosures in which the animals were difficult to see.[33]

DESIGNING MOBILE DIGITAL INTERPRETATION
THAT WORKS FOR VISITORS

Applying digital technology to museum settings is complex and prone to many pitfalls since visitors often behave in surprising ways, and how visitors use exhibitions is often at odds with what museum staff want or expect.[34] As many research studies have shown, the way in which visitors perceive and use museum exhibits is profoundly influenced by their prior knowledge and experience, motivations, and expectations, as Falk and Dierking point out in chapter 2 and other research supports.[35]

The operation of any mobile digital device in a museum setting can be thought of as a complex network of interactions (figure 3.1). To deliver the aims of enhancing visitors' enjoyment of and learning from the museum's collections, the handheld device must facilitate interactions among itself, its user, their companions and any handheld devices they have, the objects, and in some circumstances other visitors. Each of these interactions needs to be taken into account in the design of the device's hardware and software.

Useful lessons for the design of handheld digital devices can be gleaned from the experience of developing fixed kiosk-based computer exhibits[36] as well as from more recent research into mobile digital technology in museum

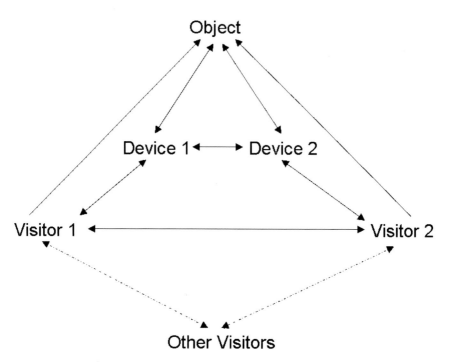

FIGURE 3.1
The design of a mobile digital device needs to take into account the complex network of interactions between the visitor and their companions, their handheld digital devices, the object, and other visitors.

settings,[37] some of which is presented in this book. These studies have demonstrated that

- visitors need to have an appropriate mental model of the digital exhibit's purpose and operation so that they can make useful predictions of how it will respond to different inputs;
- visitors need to be able to quickly understand how to operate the device; and
- the device needs to dovetail with the activity of museum visiting—that is, it does not interfere with visitors' interactions with other people or exhibits; it is available as soon as it is required and is unobtrusive when it is not needed.

THE NEED FOR AN APPROPRIATE MENTAL MODEL

The design of any piece of equipment should provide its users with a useful mental model of how it operates.[38] The users' mental model needs to allow them to make accurate predictions regarding the response of the device—for example, why it has responded in a particular way, or how it will respond to a particular input. It does not necessarily matter if the users' mental model fails to correspond exactly to the actual operation of the device, provided that the users' predictions broadly match what happens. However, where users' mental model leads them to draw inappropriate inferences, difficulties are certain to arise.

For example, observation of visitors at computer exhibits and with mobile digital technology has shown that they frequently assume any change on the screen is caused by touching it.[39] While this is an appropriate and useful mental model for a touchscreen interface, it inevitably leads to problems where the image is changing automatically or because of an input from another visitor using a different interface control, such as a push button.

Visitors also need an appropriate mental model of the device's purpose. One recent study identified three different possible functions for computer exhibits—introducing visitors to an exhibition; providing complementary information about a particular exhibit (i.e., acting as an object label); or providing supplementary information (e.g., an interactive educational game). Studying the degree of success visitors had in engaging with the computer exhibits, it was found that they need to be able to quickly identify the intended function and target audience, and that they often make assumptions about these based on subtle clues such as the location and physical orientation of the screen, its position in relation to the rest of the exhibition, as well as the title, design, and content of the first few screens. However, the clues provided were found not always to convey a useful impression of the exhibit's intended purpose or audience.[40]

Problems arising from inappropriate and unhelpful mental models have also been found in various applications of mobile digital technology. Silvia Filippini-Fantoni and J. P. Bowen recently investigated why there was such a disappointingly low usage of "bookmarking" functions on computer-kiosk and mobile digital devices in various museums. From interviews with visitors, they found that users thought the information they tagged would only be stored on the computer in the museum and hence would not be accessible to them once they left the building. This misconception was caused by the use of the term "bookmark," which failed to convey an appropriate model of what the system would actually do. It was found instead that using the term "e-mail" provided a much more useful mental model of how the system worked and successfully encouraged visitors to use these functions.[41]

Similarly, confusion arose from use of the term "texting" in the prototype version of Tate Modern's PDA tour.[42] In this case, visitors expected the text messaging function to work exactly like that found on a mobile phone and were disappointed when they found they could only select from a limited menu of preset messages.

An evaluation of a digital audio tour at the Science Museum, London, was conducted to determine why the take-up rate was much lower than expected. This study showed that many visitors seemed to harbor a misconception about the operation of the audio guide. In a survey of one hundred visitors who had not chosen to purchase the audio guide, the most common reason given was that they did not wish to follow a set route. However, the objects featured on the tour were numbered, the audio guide had a random access memory, and visitors could obtain information about objects in any order they wished. The perception among visitors was nonetheless that they would be forced to follow the route in numerical order. Intriguingly, even visitors who used the audio guide shared this negative impression.[43] Such prejudicial preconceptions were also found among visitors before they used one of the PDA tours at Tate Modern.[44]

Providing visitors with control over the type, amount, and timing of information has been shown to be key to the success of any mobile digital technology.[45] Yet, as these case studies illustrate, simply providing this functionality is not enough; it is also vital that the user is made aware that they have this degree of control.

Designing technology to mirror the operation of existing technology with which visitors are already familiar makes such devices much easier to operate, since they can use existing mental models.[46] For example, visitors found Tate Modern's PDA tour easy to operate because its screen layout used standard Web conventions with which they were already familiar.[47] However, as the above examples illustrate, such metaphors can be equally unhelpful when they convey inappropriate expectations. Often the only way to determine whether the analogies used in the design of a device are useful or not is through the careful testing of prototypes with real visitors in real museum settings.[48]

UNDERSTANDING HOW TO OPERATE THE DEVICE

Even if the visitors have an appropriate mental model of how the digital device functions, they still need to understand how to enter the necessary inputs. While considerable research has been conducted into designing usable computer software and hardware in work and home settings, relatively little such research has been conducted in museums. What research has been done has shown that visitors' use of computers in museums is often very different from that in other settings and presents the designer of a digital museum exhibit with unique problems to address. From the testing of over three hundred prototype digital exhibits at the Science Museum in London, a consistent pattern of visitor needs has emerged.[49] While these exhibits were fixed, kiosk-based computers, useful lessons can still be drawn for the design of mobile digital devices for museum settings:

- The pacing of content, instructions, and activities needs to be much faster than would be found on a conventional computer game or website. A study of over one hundred museums found that visitors typically spend less than twenty minutes in an exhibition, regardless of its size or topic.[50] Visitors thus only allocate a few minutes to an interactive exhibit, so instructions and conclusions need to be brief and preferably integrated into the main experience. Even for a mobile digital device that visitors carry with them throughout their visit, the time available for them to learn how to operate it will be dramatically less than would be the case for a computer game or website.
- Visitors very rarely use "help" buttons because they expect such options to provide lengthy, boring, and complex text-based information.
- Educational content in interactive games must be incorporated throughout the activity and not relegated to the concluding screens, which visitors largely ignore.
- At database-style exhibits (such as computer labels or digital guided tours) visitors will be either browsing to gain an overview of the content or searching to find specific information. Visitors often switch back and forth between these two modes, so the interface needs to accommodate both types of use. The main navigation buttons used by visitors are the forward, back, and home page buttons. More complex search systems are greatly appreciated but are used to a much lesser extent.

- Touchscreen interfaces work well, provided that visitors can clearly identify which parts of the screen are active. Visitors will often assume that anything moving or prominent on the screen is active. Due to the lack of tactile feedback, visitors often find it difficult to determine whether they have successfully activated a control. It is therefore important to provide both visual and auditory feedback when an on-screen button has been pressed.
- Visitors expect computer exhibits to respond instantly. Any delay is assumed to be caused either by them not pressing a button hard enough or because the computer is broken.

Prototype testing of mobile digital technology in museum settings has identified other problems unique to this particular setting. A frequent source of problems for users of mobile digital guides is the difficulty of aligning the virtual content of the digital tour with what the visitor is actually experiencing at that moment in the real world. Visitors are often frustrated by not being able to find a particular artwork or object featured on the virtual tour or, vice versa, not being able to locate information about the artwork or object that they are currently viewing.[51] In a PDA-based multiplayer game, children were observed to experience problems when there was a disjunction between the real environment they were moving about in and the virtual world on the PDA screen.[52] In all of these studies it was found that visitors' enjoyment and use of the mobile device was severely degraded when there was not a seamless match between the content on the handheld device and the real exhibit experience.

A project involving handheld digital devices in a historic house overcame this problem by using location-sensitive technology, which presented visitors with images of the objects that they were currently facing. To find out about an object, visitors simply had to touch the corresponding image on the screen.[53] Another successful technique to match the virtual and real experience was used in Tate Modern's PDA tour: audio prompts were used to alert visitors to changes on the PDA screen that they might not have been looking at and inform them of when to switch their attention back to the artwork.[54]

DOES DIGITAL TECHNOLOGY DISRUPT SOCIAL INTERACTION?

In *The Myth of the Paperless Office*, Abigail Sellen and Richard Harper pose an intriguing question: why, despite the enormous increase in the availability and power of digital technology, have paper-based working practices persisted and the amount of paper that is used increased? The authors explore a range of case studies, including an air traffic control center, a local police force, and the purchasing department of a chocolate manufacturer, in which attempts were made to replace traditional paper-based systems with digital technology. The authors found that the traditional paper-based systems afforded certain ways of working that the digital system struggled to match. Paper-based systems were often easier to organize, annotate, and personalize than computer files, and in particular far less disruptive of the social interactions that were vital to successful working practices.

Sellen and Harper's findings illustrate that the successful application of digital technology needs to be guided by a detailed understanding of how people use existing systems and what types of activities these systems afford. In particular, the design of digital technology needs, at the very least, to allow existing social interaction to continue, something to which the computer screen is not well suited.[55]

The successful application of digital technology to museums similarly requires that it fit within the complex network of social interactions between users, their companions, other visitors, and any staff member that they come into contact with. It must also coordinate with the personal interests, motivations, prior knowledge, and learning styles of the individual user, and be suited to the physical environment of the museum—the structure of the building, the layout of exhibitions, the design of the displays, and their content. In other words, any mobile digital technology needs to be able to fit within Falk and Dierking's Contextual Model of (Museum) Learning—the complex interplay between the physical, personal, and sociocultural context of the museum experience, which they detail in chapter 2.

Designing digital technology to fit within the social context is particularly important, given that the museum experience is an inherently social activity. Most people visit museums as part of a group—a family, school party, or adult couple—and spending time with friends and family has been found to be one of the key motivations for visiting.[56] Furthermore, social interaction among visitors (particularly in the form of conversation) is now widely recognized as being crucial to the process of learning.[57] Exhibits that disrupt this social interaction have been shown to be highly detrimental to visitors' learning.[58]

Concern has been expressed that digital technology—both in the form of fixed, kiosk-based computer exhibits and mobile devices—disrupts social interaction between visitors and, in particular, that it inhibits conversation. Studies have shown, for example, that there is less conversation and other forms of social interaction at kiosk-based computer exhibits,[59] while other research indicates that handheld devices can create a sense of isolation among visitors.[60]

The risk of mobile digital technology disrupting conversation seems to be particularly acute when headphones are used, since visitors have been found to experience difficulties coordinating with their companions—including their movement through the exhibition and their choices of information, not to mention the difficulties of hearing what they say.[61]

However, the evidence for the impact of digital technology upon social interaction is equivocal. Other recent research has shown that the use of digital technology does not inevitably lead to reduced social interaction and conversation and, if appropriately designed, can actually facilitate them. Many studies have shown that fixed, kiosk-based computer exhibits are typically used by two or three visitors at a time,[62] providing a shared experience that can elicit mutually supportive social interaction.[63] Digital interpretation of one interactive exhibit actually increased social interaction and conversation between parents and children.[64] Another study revealed more process-focused conversation around an interactive computer exhibit about pollination than at the neighboring live plant.[65] The amplified scale, reduced complexity, and ability to repeat the experience provided by the interactive computer exhibit afforded more opportunities for such conversation.

As with kiosk-based computer exhibits, it has been shown that properly designed and appropriately applied mobile technology can actually be an aid, rather than a barrier, to conversation and social interaction. The innovative design of a digital handheld guide for a historic house illustrates how such technology can effectively dovetail with the complex social interaction between visitors.[66] In this case, the researchers initially developed handheld digital devices that delivered content via speaker instead of headphones. This meant visitors could share information, coordinate their movements around the space, and engage in conversations about what they had discovered.[67] Acknowledging that such a system would not be practical in a busy gallery, the research team then developed a handheld device with single earpiece headsets, with which visitors could either select their own audio track or eavesdrop on that of their companions via a wireless network.[68]

This system thus provided visitors with control over what to listen to and freedom to move independently around the exhibition, while also allowing a joint experience with their companions through a "shared audio space." The content of the mobile guide was also divided into sections of six to twenty-seven seconds in length—a much shorter duration than is normally found on audio guides—in order to allow plenty of time for conversation. The combination of these design features was found to be successful in encouraging richer and more extensive interaction among visitors than was the case with either a more conventional digital handheld guide or a paper-based guidebook. The researchers propose that the design of the digital handheld guide was successful because it allowed visitors to manage the competing demands upon their attention from the objects, the digital guide, and their companions, while still providing them with control over what they experienced.[69]

MATCHING THE NEEDS AND WANTS OF INDIVIDUAL VISITORS

As well as the social context, the nature of the museum experience will be profoundly influenced by the personal context of each visitor's individual motivations, prior knowledge, and experience, as in Falk and Dierking's model. As Sellen and Harper found in a wide variety of workplace settings,[70] digital technology that has been designed with little consideration of users' preferences and working practices is unlikely to be adopted. Other research has shown that the low level of adoption of the Internet by certain segments of society can be caused as much by its perceived lack of relevance to their personal interests as by a lack of access to the technology and the knowledge of how to use it.[71]

Mobile digital technology in museums therefore needs to be able to cater to a diverse range of needs and wants, especially as visitors will be engaging with it for much longer periods of time than they would with a static label, a kiosk-based computer exhibit, or even a live interpreter.

However, one of the biggest potential problems faced by developers of mobile digital technology for museums may actually be an embarrassment of riches. It is possible to include such a vast array of information, resources, and activities that the designer may be tempted to try and cater to every possible audience. Yet research has shown that technology that attempts to be "all things to all visitors" is likely to fail because the visitors are overwhelmed by choice—much of which is of little interest or use to them.

One study identified five possible functions of a mobile digital device in a science center setting: providing detailed information about exhibits, helping visitors to communicate with other people, recording information for later reference, guiding visitors around exhibits, and providing suggestions for how to operate exhibits. The findings suggest that the combination of all five functions was too complex, and for many users the additional content, while welcome, was unnecessary in a hands-on situation. The research concluded that a device's most useful function in a science center is to record information for later reference.[72]

Another problem is that the different needs and wants of different types of visitors are sometimes incompatible. Catering to one type of audience may exclude other audiences. Research has identified four distinct categories of satisfying experience among museum visitors: object focused, cognitive (i.e., informational), introspective, and social. It found that there were inevitable conflicts between experiences that satisfied object-focused motivations and those that satisfied cognitive motivations, as well as conflicts between those that satisfied introspective motivations and those that satisfied social motivations.[73]

It may therefore be more effective to target a limited range of options to a more precisely defined audience with particular needs and wants. Filippini-Fantoni and Bowen propose that the apparent failure of various digital exhibits is actually due to them not being appropriately targeted at specific audiences. While many visitors, for example, were not interested in "bookmarking" information on museum websites, computer kiosk exhibits, or PDA tours, certain types of visitors—teachers, researchers, and subject enthusiasts—did find these facilities to be particularly useful.[74] Similarly, describing a PDA guide in chapter 8 of this book, Sherry Hsi suggests that, while the system did not fit well with the needs and wants of a family or adult group, for teachers the ability to access extra information, as well as capture and recall elements of the experience after the visit, was immensely valuable.

COPING WITH THE PHYSICAL CONTEXT OF THE VISIT

The design of mobile digital technology for museums needs to take account of the unusual physical environment into which it will be placed, according to Falk and Dierking in chapter 2. It needs to account for the large number of people and other electronic equipment within the space, the speed at which visitors use museum interpretation, the number of distracting stimuli that visitors need to cope with, the density of objects on display, the often nonintuitive arrangement of exhibitions, and many other factors.

Different types of museums have very different approaches to organizing content and exhibits, so it is likely that any particular system needs to be specially designed for its intended location. In larger spaces, it can be harder to navigate and identify individual exhibits; for example, in a large science center, users tend to identify exhibits spatially (i.e., as "that exhibit over there"), rather than by name or description, and mobile digital guides need to be designed to take this into account.[75] Visitors have often expressed frustration when the virtual content of a PDA or audio tour does not closely match the actual content of the exhibition, such as when there are artworks or objects not featured on the tour but that the visitors wanted information about.[76]

Specific types of museum experiences can also throw up specific challenges for developers. For example, one of the challenges of using a PDA in a science center is the need for visitors to have both hands free to operate the exhibits, according to Hsi and others, something that holding a PDA does not facilitate.[77]

As with the previously discussed social and personal barriers, the physical barriers to the successful application of digital technology in museums are unpredictable and location specific, and there is no substitute for careful testing of the content, hardware, and software of digital technology with real visitors in real museum settings.

CHARACTERISTICS OF SUCCESSFUL MOBILE DIGITAL TECHNOLOGY

Reviewing the research into visitors' needs, wants, expectations, and use of digital technology in museum settings, it is possible to provide some general guidelines for successful mobile digital interpretation:

1. One size does not fit all: different audiences have different needs and wants that may not be compatible within a single device. While mobile digital technology can deliver many different types of experience, a single device should not attempt to cater to all possible needs and wants, since this is likely to confuse and overwhelm users. Instead, a target audience should be identified for which the mobile digital device is likely to be useful and accessible.

2. Exploit visitors' existing knowledge and mental models: the mobile digital technology should behave, or at least seem to behave, like other forms of technology with which the visitor is familiar.

3. Make people aware that they have control over the device: it is not enough just to provide a flexible system; people also need to be aware that they have control over the flow of information to suit their immediate needs.

4. There needs to be a seamless match between what the visitor experiences in the real world and the information provided via the mobile digital device. Visitors need to realize that the two sources of information correspond to one another, and be able to make the necessary adjustments if they no longer align.

5. Design content and hardware for social interaction: audio and visual information must be shareable with other people; information should be provided in small chunks to allow plenty of time for conversation and viewing of the object or artwork; visitors need to know what their companions are seeing or hearing; and conversation must be possible while visitors are using the device.

6. Prototype versions of the device must be tested with representative samples of the target audience, in realistic settings, to ensure that they can quickly learn how to use the controls. Even when the device is based on familiar technology, the novelty of the museum setting can result in unexpected patterns of use. Prototype testing should focus on determining what visitors think the purpose of the device is and how they believe it operates.

NOTES

1. The term "computer kiosk exhibits" refers to computer terminals that are installed in permanent housings at a fixed location in the museum.

2. Mobile handheld devices include personal digital assistants (PDAs), mobile phones, and digital audio guides.

3. Richard Cassels, "Learning Styles," in *Developing Museum Exhibitions for Lifelong Learning*, ed. Gail Durbin (London: The Stationery Office on behalf of the Group for Education in Museums, 1996), 38–45; Morna Hinton, "The Victoria and Albert Museum Silver Galleries II: Learning Style and Interpretation Preference in the Discovery Area," *Museum Management and Curatorship* 17, no. 3 (1998): 253–94.

4. Sue Allen, *Finding Significance* (San Francisco: Exploratorium, 2004).

5. David Judd, Morna Hinton, and Frances Lloyd-Baynes, "Interpretation in the Galleries," in *Creating the British Galleries at the V&A: A Study in Museology*, ed. Christopher Wilk and Nick Humphrey (London: Victoria and Albert Museum Publications, 2004), 145–63.

6. Susie Fisher, "Multimedia and BSL Tours at Tate Modern: Stage 2 Qualitative Research" (unpublished evaluation report by the Susie Fisher Group for Tate Modern, London, February 2004).

7. David Williamson-Shaffer and Mitchel Resnick, "'Thick' Authenticity: New Media and Authentic Learning," *Journal of Interactive Learning Research* 10, no. 2 (1999): 195–215.

8. Reed Stevens and Roger Hall, "Seeing Tornado: How Video Traces Mediate Visitor Understandings of (Natural?) Phenomena in a Science Museum," *Science Education* 81, no. 6 (1997): 735–47.

9. Science Museum, London, "Energy—Fueling the Future," at www.science museum.org.uk/on-line/energy/site/about.asp (accessed May 27, 2007).

10. Sherry Hsi, "A Study of User Experiences Mediated by Nomadic Web Content in a Museum," *Journal of Computer Assisted Learning* 19, no. 3 (2003): 308–19; Margaret Fleck, Marco Frid, Tim Kindberg, Rakhi Rajani, Eamonn O'Brien-Strain, and Mirjana Spasojevic, "From Informing to Remembering: Deploying a Ubiquitous System in an Interactive Science Museum," *Pervasive Computing* 1, no. 2 (2002): 13–21; Silvia Filippini-Fantoni and Jonathan Bowen, "Bookmarking in Museums: Extending the Museum Experience beyond the Visit?" In *Museums and the Web 2007: Proceedings*, ed. Jennifer Trant and David Bearman. (Toronto: Archives and Museum Informatics, 2007), at www.archimuse.com/mw2007/papers/filippini-fantoni/filippini-fantoni.html (accessed August 13, 2007).

11. Steve Bitgood, "What Do We Know about School Field Trips?" in *What Research Says about Learning in Science Museums*, vol. 2 (Washington, D.C.: Association of Science-Technology Centers, 1993), 12–16; David Anderson, Keith B. Lucas, Ian S. Ginns, and Lynn D. Dierking, "Development of Knowledge about Electricity and Magnetism during a Visit to a Science Museum and Related Post-Visit Activities," *Science Education* 84, no. 5 (2000): 658–79; Jeanette Griffin, "Research on Students and Museums: Looking More Closely at the Students in School Groups," *Science Education* 88, suppl. I (2004): S59–S70; Tina Jarvis and Anthony Pell, "Factors Influencing Elementary School Children's Attitudes Towards Science before, during and after a Visit to the UK National Space Centre," *Journal of Research in Science Teaching* 42, no. 1 (2005): 53–83.

12. Fisher, "Multimedia and BSL Tours"; Filippini-Fantoni and Bowen, "Bookmarking in Museums"; Keri Facer, Richard Joiner, Danaë Stanton, Josephine Reid, Richard Hull, and David S. Kirk, "Savannah: Mobile Gaming and Learning?" *Journal of Computer Assisted Learning* 20, no. 6 (2004): 399–409; Heleen Van Loon, Kris Gabriël, Kris Luyten, Daniel Teunkens, Karel Robert, Karin Coninx, and Elke Manshoven, "Supporting Social Interaction: A Collaborative Trading Game on a PDA," in *Museums and the Web 2007: Proceedings*, ed. Jennifer Trant and David Bearman (Toronto: Archives and Museum Informatics, 2007), at www.archimuse.com/mw2007/papers/vanLoon/vanLoon.html (accessed August 13, 2007).

13. Laura Naismith, Peter Lonsdale, Giasemi Vavoula, and Mike Sharples, *Literature Review in Mobile Technologies and Learning* (Bristol, UK: Futurelab, 2005), at www.futurelab.org.uk/research/reviews/reviews_11_and12/11_01.htm (accessed May 21, 2007).

14. For a review of recent projects, see Fern Faux, Angela McFarlane, Neil Rocho, and Keri Facer, *Learning with Handheld Technologies: A Handbook from Futurelab* (Bristol, UK: Futurelab, 2006), at www.futurelab.org.uk/research/handbooks/05_01.htm (accessed May 22, 2007); Filippini-Fantoni and Bowen, "Bookmarking in Museums."

15. Beverly Serrell and Britt Raphling, "Computers on the Exhibit Floor," *Curator* 35, no. 3 (1992): 181–89; see also discussion in Judd et al., "Interpretation in the Galleries."

16. Creative Research, "Genetic Choices? Results of a Visitor Evaluation" (unpublished evaluation report by Creative Research produced for the Science Museum, London, April 1997); Creative Research, "Evaluation of *Future Foods?*" (unpublished evaluation report produced for the Science Museum, London, March 1998); Eric D. Gyllenhaal and Deborah L. Perry, "Doing Something about the Weather: Summative Evaluation of Science Museum of Minnesota's *Atmospheric Explorations* Computer Interactives," *Current Trends in Audience Research and Evaluation* 11 (1998), www.selindaresearch.com/GyllenhaalAndPerry1998Weather.pdf (accessed November 4, 2007); Ben Gammon, "Visitors' Use of Computer Exhibits," *Informal Learning Review* 38 (1999): 10–13; Cody Sandifer, "Technological Novelty and Open-Endedness: Two Characteristics of Interactive Exhibits that Contribute to the Holding of Visitor Attention in a Science Museum," *Journal of Research in Science Teaching* 40, no. 2 (2003): 121–37; Robin Meisner, Dirk vom Lehn, Christian Heath, Alexandra Burch, Ben Gammon, and Molly Reisman, "Exhibiting Performance: Co-participation in Science Centres and Museums," *International Journal of Science Education* 29, no. 12 (2007): 1531–35.

17. Cody Sandifer, "Technological Novelty and Open-Endedness: Two Characteristics of Interactive Exhibits that Contribute to the Holding of Visitor Attention in a Science Museum," *Journal of Research in Science Teaching* 40, no. 2 (2003): 121–37.

18. Creative Research, "Evaluation of *Future Foods?*; Creative Research, "Genetic Choices?"

19. Victoria and Albert Museum, "British Galleries," at www.vam.ac.uk/collections/british_galls/index.html (accessed May 27, 2007).

20. Creative Research, *Summative Evaluation of the British Galleries* (evaluation report produced for the Victoria and Albert Museum, London, September 2002), at www.vam.ac.uk/files/file_upload/5874_file.pdf (accessed December 21, 2007).

21. Morris Hargreaves McIntyre, *Engaging or Distracting: Visitor Responses to the Interactives in the V&A British Galleries* (evaluation report for the Victoria and Albert Museum, London, 2003), at www.vam.ac.uk/files/file_upload/5877_file.pdf (accessed May 21, 2007).

22. Cognitive Applications, *The Micro Gallery: A Survey of Visitors* (evaluation report for the National Gallery, London, December 1992), at www2.cogapp.com/homeimg/Micro%20Gallery%20Survey%20Report.pdf (accessed December 21, 2007).

23. Nancy Proctor, Jane Burton, and Chris Tellis, "The State of the Art in Museum Handhelds in 2003," in *Museums and the Web 2003: Selected Papers from an International Conference* (Charlotte, N.C., March 2003), at www.archimuse.com/mw2003/papers/proctor.html (accessed April 2007); Nancy Proctor and Jane Burton, "Tate Modern Multimedia Tour Pilots 2002–2003 (unpublished evaluation report from Tate Modern, London, 2003); Fisher, "Multimedia and BSL Tours"; TWResearch, "Evaluation of a Multimedia Guide Accompanying the Frida Kahlo Exhibition" (unpublished evaluation report for Tate Modern, London, 2005).

24. Fleck et al., "From Informing to Remembering"; Dirk vom Lehn and Christian Heath, "Displacing the Object: Mobile Technologies and Interpretative Resources" (paper presented at ICHIM 03, Paris, September 2003), at www.ichim.org/ichim03/PDF/088C.pdf (accessed May 22, 2007).

25. Science Museum, London, "Energy Hall," at www.sciencemuseum.org.uk/visitmuseum/galleries/energy_hall.aspx (accessed August 9, 2007).

26. Morris Hargreaves McIntyre, "CIPs in Context: Understanding Visitor Usage of Computer Information Points at the Science Museum" (unpublished evaluation report for the Science Museum, London, 2006).

27. Science Museum, London, "Nanotechnology—Small Science, Big Deal," at www.sciencemuseum.org.uk/antenna/nano/ (accessed August 9, 2007).

28. Science Museum, London, "Nanotechnology—Small Science, Big Deal: Summative Evaluation" (unpublished evaluation report, 2005).

29. McIntyre, "CIPs in Context."

30. Allison Woodruff, Paul M. Aoki, Amy Hurst, and Margaret H. Szymanski, "Electronic Guidebooks and Visitor Attention," in *Proceedings of 6th International Cultural Heritage Informatics Meeting*, ed. David Bearman and Jennifer Trant (Philadelphia: Archive and Museum Informatics, 2001), 437–54.

31. Paul M. Aoki, Rebecca E. Grinter, Amy Hurst, Margaret H. Szymanski, James D. Thornton, and Allison Woodruff, "Sotto Voce: Exploring the Interplay of Conversation and Mobile Audio Space," in *Proceedings of the SIGCHI Conference on Human Factors in Computing Systems: Changing Our World, Changing Ourselves* (New York: ACM Press, 2002), 431–38; Allison Woodruff, Paul M. Aoki, Rebecca E. Grinter, Amy Hurst, Margaret H. Szymanski, and James D. Thornton, "Eavesdropping on Electronic Guidebooks: Observing Learning Resources in Shared Listening Environments," in *Museums and the Web 2002: Selected Papers from an International Conference*, ed. David Bearman and Jennifer Trant (Pittsburgh: Archives and Museum Informatics, 2002), 21–30.

32. Proctor, Burton, and Tellis, "The State of the Art"; Fisher, "Multimedia and BSL Tours"; Gillian Wilson, "Multimedia Tour Programme at Tate Modern" (paper presented at the Museums and the Web Conference, Washington D.C., April 2004), at www.archimuse.com/mw2004/papers/wilson/wilson.html (accessed May 27, 2007); TWResearch, "Evaluation of a Multimedia Guide."

33. Luciana Baptista, "BBC Collect: Stapler Trial at London Zoo: Preliminary Findings" (unpublished evaluation report for the Zoological Society of London, November 2005).

34. Gammon, "Visitors' Use of Computer Exhibits"; Benjamin Gammon, "Everything We Currently Know about Making Visitor-Friendly Mechanical Interactive Exhibits," *Informal Learning Review* 39 (1999): 1–13; Sharon MacDonald, *Behind the Scenes at the Science Museum* (New York: Berg, 2002); Sue Allen and

Joshua Gutwill, "Designing with Multiple Interactives: Five Common Pitfalls," *Curator* 47, no. 2 (2004): 199–212.

35. Zahava D. Doering, Andrew J. Pekarik, and Audrey E. Kindlon, "Exhibitions and Expectations: The Case of 'Degenerate Art,'" *Curator* 40, no. 2 (1997): 126–41; Zahava D. Doering, "Strangers, Guests or Clients? Visitor Experiences in Museums," *Curator* 42, no. 2 (1999): 74–87; Andrew J. Pekarik, Zahava D. Doering, and Adam Bickford, "Visitors' Role in an Exhibition Debate: Science in American Life," *Curator* 42, no. 2 (1999): 117–29; John H. Falk and Lynn D. Dierking, *Learning from Museums: Visitor Experiences and the Making of Meaning* (Walnut Creek, Calif.: AltaMira, 2000); MacDonald, *Behind the Scenes.*

36. Gyllenhaal and Perry, "Doing Something about the Weather"; Gammon, "Visitors' Use of Computer Exhibits"; Hsi, "A Study of User Experiences"; Allen and Gutwill, "Designing with Multiple Interactives"; McIntyre, "CIPs in Context."

37. Aoki et al., "Sotto Voce"; Woodruff et al., "Eavesdropping on Electronic Guidebooks"; Hsi, "A Study of User Experiences"; Fisher, "Multimedia and BSL Tours"; Filippini-Fantoni and Bowen, "Bookmarking in Museums."

38. Donald A. Norman, *The Design of Everyday Things* (New York: Basic Books, 2002).

39. Gammon, "Visitors' Use of Computer Exhibits"; Facer et al., "Savannah."

40. McIntyre, "CIPs in Context."

41. Filippini-Fantoni and Bowen, "Bookmarking in Museums."

42. Fisher, "Multimedia and BSL Tours."

43. Science Museum, London, "Soundbytes Audio-Tour, Summative Evaluation (unpublished evaluation report, 2003).

44. TWResearch, "Evaluation of a Multimedia Guide."

45. Allison Woodruff, Paul M. Aoki, Amy Hurst, and Margaret H. Szymanski, "Electronic Guidebooks and Visitor Attention," in *Proceedings of 6th International Cultural Heritage Informatics Meeting*, ed. David Bearman and Jennifer Trant (Philadelphia: Archive and Museum Informatics, 2001), 437–54; Science Museum, "Soundbytes Audio-Tour"; TWResearch, "Evaluation of a Multimedia Guide."

46. Norman, *The Design of Everyday Things.*

47. Fisher, "Multimedia and BSL Tours."

48. Gammon, "Visitors' Use of Computer Exhibits"; Gammon, "Everything we Currently Know"; Allen and Gutwill, "Designing with Multiple Interactives."

49. Gammon, "Visitors' Use of Computer Exhibits."

50. Beverly Serrell, "Paying Attention: The Duration and Allocation of Visitors' Time in Museum Exhibitions," *Curator* 40, no. 2 (1997): 108–25.

51. Woodruff et al., "Electronic Guidebooks and Visitor Attention"; Hsi, "A Study of User Experiences"; Fisher, "Multimedia and BSL Tours"; TWResearch, "Evaluation of a Multimedia Guide."

52. Facer et al., "Savannah."

53. Woodruff et al., "Electronic Guidebooks and Visitor Attention."

54. Fisher, "Multimedia and BSL Tours"; Wilson, "Multimedia Tour Programme."

55. Abigail J. Sellen and Richard H. Harper, *The Myth of the Paperless Office* (Cambridge, Mass.: MIT Press, 2002).

56. Andrew J. Pekarik, Zahava D. Doering, and David A. Karns, "Exploring Satisfying Experiences in Museums," *Curator* 42, no. 2 (1999): 152–73; MacDonald, *Behind the Scenes.*

57. Falk and Dierking, *Learning from Museums*; Sue Allen, "Looking for Learning in Visitor Talk: A Methodological Exploration," in *Learning Conversations in Museums*, ed. Gaea Leinhardt, Kevin Crowley, and Karen Knutson (Mahwah, N.J.: Erlbaum, 2002), 259–303.

58. Rochel Gelman, Christine M. Massey, and Mary McManus, "Characterizing Supporting Environments for Cognitive Development: Lessons from Children in a Museum," in *Perspectives on Socially Shared Cognition*, ed. Lauren B. Resnick, John M. Levine, and Stephanie D. Teasley (Washington, D.C.: American Psychological Association, 1991), 226–56; Kevin Crowley and Maureen Callanan, "Describing and Supporting Collaborative Scientific Thinking in Parent-Child Interactions," *Journal of Museum Education* 23, no. 1 (1998): 12–17; Allen and Gutwill, "Designing with Multiple Interactives."

59. Gyllenhaal and Perry, "Doing Something about the Weather"; vom Lehn and Heath, "Displacing the Object"; Christian Heath, Dirk vom Lehn, and Jonathan Osborne, "Interaction and Interactives: Collaboration and Participation with Computer-Based Exhibits," *Public Understanding of Science* 14, no. 1 (2005): 91–101.

60. Hsi, "A Study of User Experiences"; Fleck et al., "From Informing to Remembering."

61. Woodruff et al., "Electronic Guidebooks and Visitor Attention"; Fisher, "Multimedia and BSL Tours."

62. Gyllenhaal and Perry, "Doing Something about the Weather"; Gammon, "Visitors' Use of Computer Exhibits"; Science Museum, London, "Nanotechnology— Small Science, Big Deal, Summative Evaluation" (unpublished evaluation report, 2005); Science Museum, London, "Energy: Fuelling the Future, Summative Evaluation" (unpublished evaluation report, 2005).

63. Meisner et al., "Exhibiting Performance."

64. Gelman et al., "Characterizing Supporting Environments."

65. Catherine Eberbach and Kevin Crowley, "From Living to Virtual: Learning from Museum Objects," *Curator* 48, no. 3 (2005): 317–38.

66. Woodruff et al., "Electronic Guidebooks and Visitor Attention"; Aoki et al., "Sotto Voce"; Woodruff et al., "Eavesdropping on Electronic Guidebooks."

67. Woodruff et al., "Electronic Guidebooks and Visitor Attention."

68. Aoki et al., "Sotto Voce"; Woodruff et al., "Electronic Guidebooks and Visitor Attention."

69. Aoki et al., "Sotto Voce"; Woodruff et al., "Electronic Guidebooks and Visitor Attention."

70. Sellen and Harper, *The Myth of the Paperless Office.*

71. Keri Facer, "What Do We Mean by the Digital Divide? Exploring the Roles of Access, Relevance and Resource Networks," in *The Digital Divided* (Coventry, UK: Becta, February 2002), at www.becta.org.uk/page_documents/research/ digidivseminar.pdf (accessed May 28, 2007).

72. Fleck et al., "From Informing to Remembering."

73. Pekarik et al., "Visitors' Role in an Exhibition Debate."

74. Filippini-Fantoni and Bowen, "Bookmarking in Museums."

75. Fleck et al., "From Informing to Remembering."

76. Science Museum, "Soundbytes Audio-Tour"; Wilson, "Multimedia Tour Programme."

77. Hsi, "A Study of User Experiences"; Fleck et al., "From Informing to Remembering."

II

DELIVERING POTENTIAL

4

Audibly Engaged: Talking the Walk

JEFFREY K. SMITH AND PABLO P. L. TINIO

Visitors enter museums with great expectations. Sometimes those expectations are clearly defined; more often, they are wholly undifferentiated.[1] Visitors cheerfully place themselves in the hands of the museum and hope for the best. In anticipation of its public, the museum arranges its displays and exhibitions, and develops material to augment the visit. One of the most ubiquitous of these augmentations is the audio tour. From the early days of linear tours replete with end-of-stop bleets and classical "traveling music," to direct-access machines with multiple options, to non-museum-sanctioned podcasts, the audio tour has become a common part of the museum experience. It has been estimated that over half of art museums offer audio tours,[2] and that they are used by over thirty-five million visitors annually.[3] Decked out in headphones and high-tech paraphernalia, audio tour users wander about pushing buttons, frowning, nodding in agreement to no one in particular, and looking for all the world like they are on aesthetic life support.

They are—to some degree at least. The audio tour allows individuals to receive insight, context, anecdotal information, history, and provenance about works of art and artifacts, or about scientific or historical displays, that enhance the museum visit. Although the technology involved has developed impressively, and quite rapidly in recent years, the fundamental idea is simple: information is orally presented to the visitor via electronic device. This chapter focuses on some of the underlying psychological and sociological issues of audio augmentation in an art museum. From the simplest of issues, such as "do visitors like audio tours?" to more complex questions such as the tension between structure and freedom in a museum visit, and the impact of audio tours on that issue, we hope to shed some light on the nature and structure of visitor reaction to audio augmentation of the visit. How do visitors respond when the audio tour allows them to "talk the walk"?

WHAT IMPACT DOES INFORMATION HAVE ON VIEWING?

To augment a museum visit is to generally provide information, but the decision by art museums to provide information is not a simple matter of cutting out a section of the exhibition program and pasting it next to the artwork. Interpretive issues such as the amount and type of information to present must first be considered. Deciding whether to present descriptive or elaborative information is a complex decision, as visitors' aesthetic experiences will be differently influenced by different types of information.[4] Although most people like to have information presented with works of art, they do not typically wish to be inundated, and their preferences depend in part on the nature of the art they are viewing.[5] For example, visitors may find abstract artworks difficult to comprehend. For such works, the decision to provide interpretive information may be more necessary than for artworks considered more accessible.[6] Interpretive information has been shown to enhance people's view of the power, personal meaning, and expressiveness of artworks.[7]

These results were based mainly on research employing textual information. However, results of research using nonauditory channels can serve as the basis for work on audio augmentation. The first question addressed in this chapter, and the one addressed most thoroughly, has to do with the impact that audio information has on art viewing. Do audio tours change the nature of how people interact with art in museums? And does this impact change according to the nature of the information and the type of art viewed?

DO VISITORS DESIRE STRUCTURE OR FREEDOM?

When an adult visitor enters a museum, or an exhibition or gallery space within a museum, he or she does so voluntarily. Museum visiting is an activity of choice, as John Falk and Lynn Dierking describe in chapter 2. Once inside the museum, the visitor encounters the museum's decisions as to how to present its collection and how to afford the visitor a variety of options and restrictions in viewing the collection. The museum may choose to present the works in a very linear fashion, with limited options for entry or exit, or it may have multiple entry points and minimal information about the works. It is at this interface between visitor and museum that a certain tension exists that dictates the outcome of the museum experience. On one side there are visitor expectations regarding the visit, informational needs, learning outcomes, and general aesthetic reactions. On the other side are the museum's approach and response to the interpretive dilemma,[8] physical layout of the artworks, and other presentational decisions intended to affect the aesthetic experience. It is along these two sides of the museum experience that a tension exists between freedom and structure for the visitors. Visitors want the freedom to interpret art as they will, but at the same time they desire the structure of some organizing schematic information provided by the museum. Multiple entry points seem a good idea, but the visitor wants to make sure that nothing important is missed.

Visitors, as will be empirically shown below, desire *both* freedom and structure. It is crucial that visitors find a comfort level in this tension, and the optimal amount and types of information made accessible to them are principal in this regard. The question of unique characteristics of individual visitors can be addressed by allowing some variability *around* the presented information and the means of accessing that information. It is in this area where audio augmentation could be at an advantage over textual information or person-led tours.

Textual information printed on museum labels is necessarily the same for all visitors and basically linear in nature. Although some museums have been creative in both the content and technology of labels, they are basically noninteractive in nature. Person-led tours, though potentially more responsive to the visitor, add the element of social interaction, which may be undesirable (or intimidating) for visitors. Audio augmentation organized in a manner that provides choices in the amount and type of information is effective in helping visitors find the comfort level between freedom and structure. Many audio programs allow visitors to choose the type and depth of information they receive about a work. Furthermore, they can request overview information of a gallery or an exhibition. This makes possible the realization of audio tours that have both an overall linear component (structure) and user-controlled elements (freedom) allowing deviations to more specific information (e.g., biographical information about the artist and interpretive information). Thus, visitors will be able to navigate back and forth between freedom and structure. In the studies presented, we look more deeply at the question of freedom and structure in the museum visit.

DO AUDIO TOURS HINDER VISITOR INTERACTION?

There is the general concern regarding the influence of the use of audio guides (and similar devices) on the interaction among museum visitors, as Ben Gammon and Alexandra Burch discuss in the previous chapter. Exhibition developers and curators may be distressed by the thought of visitors walking to the beat of their personal headsets, not partaking in art-related conversations with acquaintances. Addressing this issue is essential because the majority of people do not visit museums alone but are typically accompanied by a spouse or a friend.[9]

Empirical research into the complex nature of social interaction in museum settings is extremely helpful. For example, one recent study showed that visitors go to museums expecting outcomes other than social interaction.[10] Our recent research on group tours at the Whitney Museum of American Art further shows that the concern about audio augmentation hindering social interaction might be overstated.[11] Most museums provide group tours that are usually led by a docent or a research fellow. It can be argued that group tours naturally promote social interaction more than audio tours or augmentation using labels. In addition, group tours allow the possibility for questioning of the tour guide regarding aspects of the exhibit or the tour.

Interestingly, results of our study provide evidence against the social nature of museum tours. Of the thirty participants interviewed after participating in a group tour (approximately fifty minutes in length), only three had posed a question to the tour guide. Furthermore, only one visitor talked to others during the tour. Taken together, these findings call for tracking and observation studies aimed at the influence of audio tours on social interaction. This is a critical question in museum research, especially as it relates to the use of technology. We shed some light on this in the data presented, but the studies presented here did not have this issue as a central question.

HOW DO AUDIO TOURS FOCUS THE ATTENTION OF VISITORS?

The act of entering a gallery and listening to another person's ideas brings up the issue of overemphasizing particular works of art while neglecting others and highlighting only specific elements within an artwork. As the studies presented below show, visitors' decisions as to what artworks to spend time on are influenced by the content of audio tours. Audio tours that are linear or are limited in the number of artworks discussed will influence the choices that visitors make regarding what objects to look at; by comparison, museums that provide more extensive collection coverage, and more user-controlled choices and nonlinear audio tours, present more options for viewers.

In terms of focusing the viewer's gaze on only certain elements of an artwork, eye-movement research is particularly powerful. Employing such methodology, research found that different information presented verbally changed people's descriptions of paintings but not the direction of eye gaze of the visitors.[12] Other research found that information about works of art can direct a viewer's attention toward certain aspects of a work of art and away from others.[13] In the research reviewed below, we examine both the question of which artworks visitors attend to and how information can direct visitors' attention within a work of art.

EMPIRICAL RESEARCH ON AUDIO TOURS

There is simply not a lot of research on audio tours in museums—art museums in particular. Most of the research that exists is "fugitive," that is, not readily available or published. It primarily exists as reports conducted for individual museums, although some is more generally available, and presented below are three such studies, each conducted at a major museum in New York City: the Metropolitan Museum of Art, the Jewish Museum, and the Whitney Museum of American Art. Each study was conducted with the goal of evaluating and improving some aspect or aspects of the tours offered by the museum. Each of the studies employs an element of experimentation in looking at the impact of the audio tours.

Study 1: The Metropolitan Museum of Art

The first study was conducted at the Metropolitan Museum of Art in New York City.[14] The motivation for the study came from a new audio program that the museum was implementing that included the permanent collection along with special exhibitions. Two samples of visitors were surveyed. The first was a group of 163 visitors who purchased the audio tour and agreed to complete a survey about their experience upon returning their audio equipment.

The second sample consisted of 109 visitors who were not going to use the audio tour but agreed to do so after having been offered it for free. This group
was broken down into those who had used audio tours before and those who
were using them for the first time. These three samples of visitors allow for
looking at contrasting information in the visitor responses. (Some of the
questions asked on the survey are presented in table 4.1 broken down by sub-
sample.)

Fifty-seven percent of the people who had never used an audio tour before
agreed with the statement "I found it more enjoyable than I expected." The

Table 4.1.   Metropolitan Museum of Art audio tour survey results.

Question	Purchasers	Free (Previous Users)	Free (First-Time Users)
How would you rate the audio tour? (Scale of 1 = poor, to 10 = excellent) (Mean)	8.4	8.1	7.7
Did the audio tour enhance your visit to the Museum? (1 = not at all, 10 = very much) (Mean)	8.5	7.9	7.6
Did you find the player easy to use? (1 = not easy, 10 = very easy) (Mean)	9.1	9.0	9.4
Did you find the player comfortable to wear? (1 = not at all, 10 = very) (Mean)	7.4	8.0	7.5
What kinds of information do you prefer on audio guide messages? (Percentage of visitors indicating each)			
Artistic technique	51	44	46
Artist's life	47	47	48
Subject matter of the work	61	51	43
Historical background and context	68	61	52
Insight into understanding the work	65	62	48
History of the object	56	40	43
Which of the following features would enhance the audio guide program? (Percentage of visitors indicating each)			
Background music*	55	37	39
Poems or literary readings related to work*	33	17	14
Quotes from the artists or critics of the time	59	55	50
Did you mostly go to the objects with audio messages? (Percentage of visitors indicating each)	59	52	48
Did you also read the labels accompanying the objects? (Percentage of visitors indicating each)	72	75	73

*Differences among groups statistically significant at p < .05 (chi square test).

109 visitors who had used the audio for free (groups 2 and 3 combined) were asked, "If you do not use audio tours, can you tell us why?" Only 4 people of these 109 indicated that the audio tour made it hard to talk with others. By far the most frequent response was "I never tried one before" (46 percent), followed by cost (21 percent).

The study showed that even people who have never used an audio tour before find the equipment easy and comfortable to use, and that the information provided enhances their visit. The audio tour was generally felt to enhance the visit, by all three groups, even for visitors who had never used an audio tour. Visitors want to hear about the history of the work and like insight to help them understand the work. It should be noted here that most suggestions for the kinds of information possible were favorably received by roughly half or more of the sample. The only significant differences among groups were found when we asked whether people would like poetry or music as part of the audio program. Here, the groups who had not purchased the tour were less likely to think that these would be good additions.

The issue of the tour being directive also comes up. About half of each group said that they mostly went to objects that had the audio messages. Finally, the issue of the tour being restrictive in that it inhibits conversation among individuals does not seem to be a major concern, even among individuals who have taken an audio tour for the first time.

Study 2: The Jewish Museum

The second study was conducted at the Jewish Museum in New York City.[15] The Jewish Museum "demonstrates how Jewish culture is reflected in art through 28,000 objects of different media, including fine arts, Judaica, and broadcast media."[16] This study was conducted as an evaluation of the efficacy of the audio tour program and to see how the program could be enhanced. The study surveyed 78 individuals who had taken the audio tour and 93 individuals who had not.

Focusing on those who took the tour, many of the results echo what was found in the study at the Met. Looking at figure 4.1, we see that issues such as history, meaning of the work, and context of the work are all important, in addition to some issues that were specific to the study, such as "an advanced tour that assumed substantial knowledge of Jewish history." Less important were issues about how the work was made, or music or poetry that might accompany the tour.

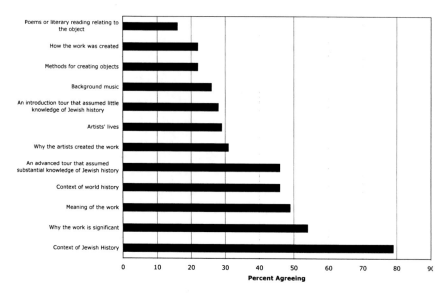

FIGURE 4.1
Preferences concerning Audio Tour Content

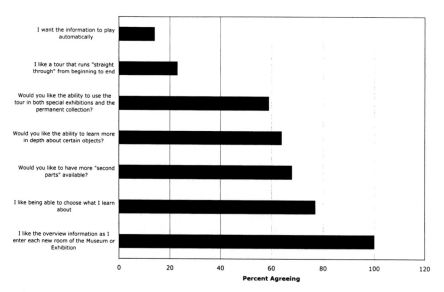

FIGURE 4.2
Preferences concerning Structure and Freedom

The Jewish Museum study also asked a number of questions that inform issues of freedom and structure for the visitor. Though the direct-access tour provided access to information about specific objects in the permanent collection, there was no "general tour" provided for visitors to follow. Instead, each room had an overview introducing the theme of the room, which automatically played as the person entered the room. This was a creative effort on the part of the museum to allow the visitor the freedom to choose particular objects, while at the same time providing some level of structure for the visit as a whole. In figure 4.2, we see that visitors appear to seek both freedom and structure. They respond positively to those aspects of the tour that provide the option to move between freedom and structure (the first five options in order), and negatively to those that restrict the movement (the last two options). Visitors want freedom; they also want structure; and they want to be able to determine how they navigate back and forth between the two.

Study 3: The Whitney Museum

The third study[17] was conducted at the Whitney Museum of American Art, "one of the world's foremost collections of twentieth-century American art."[18] The purpose of the study was to look at the impact of the museum's approach to providing audio tour information in a controlled, in situ experimental fashion. Two parts of this extensive study are presented here. The first part was taken from a survey conducted on 181 visitors who took a free audio tour of the museum; the responses are presented in figure 4.3.

What can be seen from the responses is that visitors were very positive about the audio tour program. It should be noted that the museum does not use descriptive labels; only basic information such as title, artist, and date are typically provided for a work. Visitors indicated that they enjoyed their visit more because of the audio tour, would get one again on their next visit, understood more about the works because of the audio tour, and spent more time in the galleries because of the audio tour. They did not find the audio tour too impersonal, nor did they indicate difficulty using the equipment. Also, they indicated that they would prefer using the audio equipment to reading label copy.

The second part of the Whitney study presented here was experimental in nature. A total of 240 visitors were randomly assigned to one of three treatment conditions in front of one of four works of art (two sculptures and two

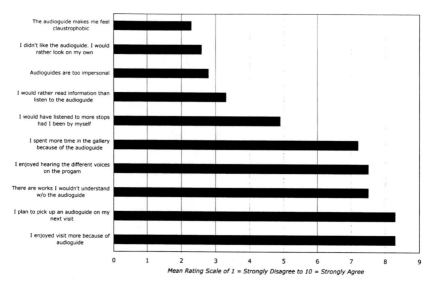

FIGURE 4.3
Ratings of Viewers Who Took Free Audio Guide

paintings). Twenty visitors completed surveys after experiencing one of the four works with one of the three groups. The four works of art were George Bellows's *Dempsey and Firpo* (1924); Arshile Gorky's *The Artist and His Mother* (ca. 1926–1936); Yayoi Kusama's *Accumulation* (ca. 1963); and Chris Burden's *America's Darker Moments* (1994). The three conditions were as follows: the museum's standard label, which consisted simply of title, artist, date, and acquisition information; a label written expressly for the study by the person in charge of exhibition information for the museum—written to represent typical art historical and stylistic information; and audio.

The audio stops varied substantially in the audio tour. At the Bellows painting, author and sports writer George Plimpton described the fighters and the fight depicted in the painting, focusing on how unpopular the champion, Tunney, was at the time. At the Gorky painting, a sociologist discussed the refugee status of the painter, who is the subject of the painting, along with his mother. The Kusama sculpture is a white chair that has been covered with dozens of phallic-shaped, white stuffed bags; the audio stop is presented by a clinical psychologist discussing an obsessive-compulsive disorder the artist suffers from. The Burden piece is a sculpture split into five pieces that represent events in American history using small metal sculptures; the audio stop was done by the sculptor himself, talking about the scenes he has depicted and why he chose them.

Visitors were approached in the galleries and asked to participate. If they agreed, they were taken to one of the four objects and subjected to one of the three groups. They stood in front of the object as long as they cared to and then were taken to a separate area to complete a short survey about what they had experienced.

One measure presented is the amount of time people spent in front of the object; the results are presented in figure 4.4. The Burden piece commanded the longest stay in all three conditions. In part, this might be attributable to the amount of information in the piece, which could be considered to be five separate works brought together as one. Watching the participants in the study, we noticed that people would often walk around the piece, contemplating each of the scenes individually. The main effect for the audio group could in part be attributable to the fact that participants listened to each of the

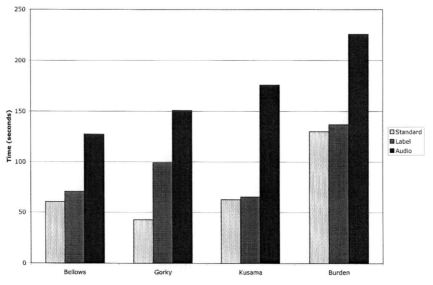

FIGURE 4.4
Mean Time Spent Looking at Artwork

audio stops all the way through. That would cause them to stay at a stop longer than what one typically finds in an art museum. In fact, the time spent in each of the three conditions was longer than typically found for a person in front of a work of art.[19] This was no doubt also due to the fact that the participants were cognizant of being in a study.

To some extent, one might argue that the *time spent* variable is artificial in that people knew they were in the study, and they had to listen to the entire audio stop. But people typically do listen to the entire audio stop. One of the positive aspects of the audio tour is that it causes people to look at works longer. Also, it keeps people's visual input focused on the work of art as opposed to diverting attention to read a label.

The second outcome is the response to the statement "I would like to see other works by this artist." The idea here was to look at the degree to which the work of art captured the interest of the participant. The results, presented in figure 4.5, show some similarities and some differences to the time results.

From figures 4.4 and 4.5, we see that the Burden work receives the longest time and the highest rating for wanting to see more works. But if we look

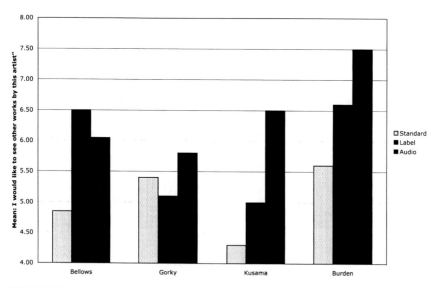

FIGURE 4.5
Mean Rating on Desire to See More Work by Artist

more closely at the graphs, we also see that it is for the Kusama work that the largest differences exist between the audio group and the other two groups; the Burden work has the second largest differences. This is particularly interesting because these are the two least readily accessible works of the four. It appears to be the case that the engaging and revealing nature of the audio stops are most effective for those works that are most difficult to understand otherwise.

The final aspect of the study at the Whitney Museum has to do with focusing attention. At each of the four objects, participants were asked a question about something that was *not* included in the audio stop. For example, at the Burden work, visitors were asked how the lighting worked for the object. At the Bellows work, people were asked about the color of the ceiling in the painting. What we found was that with each of the four objects, participants in the standard or label groups were better able to answer the question than people in the audio condition (that did not discuss the content of the question). It appears that attention was directed toward certain aspects of the work and away from others.

CONCLUSION

Visitors do not typically come to a museum or exhibition with a predetermined notion of what they are going to do, except in special cases where they know a lot about the works and what they want to do.[20] Instead, they put themselves in the hands of the museum. They look to the museum to provide a certain level of structure to their visit. They want information about the works they are to see, as well as information that helps them coordinate and organize the groupings of artwork they are viewing.

The three studies presented here show that visitors want information and that the information provided by audio tours enhances the museum experience more than other presentation formats. The study at the Met clearly shows that the majority of visitors respond well to the audio program when it is made available to them. Even individuals who had never tried one before were positive about the experience. The level of enhancement of experience may be most pronounced for artworks that are difficult to understand, such as the Burden and Kusama sculptures in the Whitney study. In this sense, information provides some structure, a cognitive schema for looking at and understanding the work.

However, it has been shown that too much structure is not optimal. Visitors also desire freedom. They do not want to be tied down to a predetermined tour that includes only selected objects. Fortunately, recent advances in audio technology allow museums to offer both freedom and structure. Visitors would like to be able to choose the objects they want information on, and they have different ideas about the type of information they want.

From a structural perspective, the Jewish Museum approach appears to be successful, providing a context that promotes resolution of the tension between freedom and structure. It achieves this by presenting an overview of the exhibition upon entry into a gallery and by offering random access to its collection of artworks. What we see here is the museum tackling the interpretive challenge head on, by allowing room for the expression of the unique qualities—whether these are background knowledge, personality characteristics, preferences, or temporary mood states and reactions—of the visitor.

Another successful expression of visitors' desire to have the museum take the lead in structuring the visit without controlling it is seen in the Whitney approach to innovative audio content combined with random access. Of particular interest here is that some stops appear to be working much better than others. In our selection of four works in this study, it appears that less accessible works need more interpretative help to be appreciated. Then again, it may be that it was something in the content that produced the observed differences, and accessibility of the works was coincidental to the preference findings. One of the areas of research we think is most critical for audio technology has to do with the nature of the messages that are presented and how they impact visitors.

The studies presented here show that audio guides do not, in the opinions of visitors, inhibit social interaction. And although they are directive to some extent, visitors do not find them constraining. The generally positive response that visitors have regarding audio guide use was consistent across studies.

So, what is the final verdict concerning audio augmentation? Let the visitors speak and they will tell you they like the audio: visitors find they enhance their visit, are easy to use and comfortable to wear, and are much better than simply reading text. This is a successful technology that continues to advance rapidly from a technological perspective. We need to continue development apace from the substantive perspective through research and innovative and creative thinking.

NOTES

1. Jeffrey K. Smith and D. W. Carr, "In Byzantium," *Curator* 44, no. 4 (2001): 335–54.

2. K. Wetterlund and S. Sayre, "2003 Art Museum Education Programs Survey," at www.museum-ed.org/ (accessed May 27, 2007).

3. Chris Tellis, "Multimedia Handhelds: One Device, Many Audiences," in *Museums and the Web 2004: Proceedings*, ed. David Bearman and Jennifer Trant (Toronto: Archives and Museum Informatics, 2004), at www.archimuse.com/mw2004/papers/tellis/tellis.html (accessed June 26, 2008).

4. K. Millis, "Making Meaning Brings Pleasure: The Influence of Titles on Aesthetic Experiences," *Emotion* 1 (2001): 320–29.

5. J. E. V. Temme, "Amount and Kind of Information in Museums: Its Effects on Visitors' Satisfaction and Appreciation of Art," *Visual Arts Research* 18 (1992): 74–81.

6. P. A. Russell and S. Milne, "Meaningfulness and Hedonic Value of Paintings: Effects of Titles," *Empirical Studies of the Arts* 15 (1997): 61–73.

7. G. C. Cupchik, L. Shereck, and S. Spiegel, "The Effects of Textual Information on Artistic Communication," *Visual Arts Research* 20 (1994): 62–78.

8. Andrew J. Pekarik, "To Explain or Not to Explain," *Curator* 47 (2004): 12–18.

9. Jeffrey K. Smith, I. Waszkielewicz, K. Potts, and B. K. Smith, "Visitors and the Audio Program: An Investigation into the Impact of the Audio Guide Program at the Whitney Museum of American Art" (unpublished report; New York: Whitney Museum of American Art, 2004).

10. R. Ballantyne and J. Packer, "Solitary vs. Shared: Exploring the Social Dimension of Museum Learning," *Curator* 48 (2005): 177–92.

11. Pablo L. Tinio, K. Potts, K., and Jeffrey K. Smith, "The Museum Tour: Visitors, Guide, and Interactivity" (unpublished manuscript, 2006).

12. M. B. Franklin, R. C. Becklen, and C. L. Doyle, "The Influence of Titles on How Paintings Are Seen," *Leonardo* 26 (1993): 103–8.

13. M. A. Gallo, "The Effect of Information on the Interpretation of Artwork (unpublished PhD diss., Rutgers University, New Brunswick, N.J., 2004).

14. Jeffrey K. Smith, "Analysis of 'Key to the Met' Audio Program" (unpublished report; New York: Metropolitan Museum of Art, 2000).

15. Jeffrey K. Smith and L. F. Smith, "Spending Time on Art," *Empirical Studies in the Arts* 19, no. 2 (2001): 229–36.

16. The Jewish Museum, "Collection Overview," at www.thejewishmuseum.org/CollectionOverview.

17. Smith et al., "Visitors and the Audio Program."

18. Whitney Museum of American Art, "History of the Whitney," at www.whitney.org/ (accessed June 26, 2008).

19 Smith and Smith, "Spending Time on Art."

20. Smith and Carr, "In Byzantium."

5

Mobile Multimedia: Reflections from Ten Years of Practice

SILVIA FILIPPINI-FANTONI AND JONATHAN P. BOWEN

In the past decade, museums have experienced a revolution with the arrival of new technologies capable of enhancing their visitors' experiences with multimedia content. Amid these new technologies, website and computer-based interactives in the gallery have augmented exhibitions to provide additional educational material. However, the Web is mainly suitable for access to information before, after, or even instead of visiting an exhibition or museum. A kiosk provides a physically fixed facility that can be used during a visit to an exhibition, but only for part of it. In contrast, a multimedia tour delivered through a handheld device allows the visitor to gain information at any point during the visit and in any order, without interfering with the aesthetics of the gallery. This means it has some unique properties that could be beneficial, if used suitably.

A multimedia tour is a guide of a museum's permanent collection, temporary exhibitions, or architecture and history, combining a variety of media including audio, text, images, videos, and occasionally also interactive programs. Because of their audiovisual nature, multimedia tours are available to visitors via screen-based handheld devices such as personal digital assistants (PDAs), iPods, and mobile phones. The level of interactivity and the supported multimedia content types vary according to the specific needs of the institution as well as the kind of device adopted. For example, iPods have mainly been used for playing videos, images, and audio, while touchscreen or keypad-based devices can be used for more complex interactions, including individual and group games. Multimedia tours are often distributed by museum staff members at the beginning of a visit, but in some cases they are available for download directly onto the visitor's personal device from the museum website (e.g., the multimedia tour of the Rembrandt exhibition at the Getty Museum in 2005).

Since the turn of the century, a variety of multimedia tours have been piloted or adopted by a range of museums worldwide. The Experience Music Project in Seattle was a pioneer, introducing in 1995 a portable, if somewhat bulky, device. In 2002, Tate Modern introduced the first wireless multimedia tour, in a three-phase pilot, leading to a commercially available tour in 2005, which since then has been sold to over 150,000 visitors. More recent examples include tours at Tate Britain, the Fitzwilliam Museum in Cambridge, the Glenbow Museum in Calgary, and the Natural History Museum in London, just to mention a few.[1]

Since 2005, the number of multimedia pilots and projects has further increased, including museums that have completely replaced their audio offerings with multimedia for both temporary exhibitions and permanent collections. At a time when a growing number of museums, especially in Europe, are moving away from simply experimenting with this technology to making it commercially available in their institutions, it is important to consider the use of these solutions within the cultural heritage sector in more detail.

Based on experience accumulated over the past ten years, in this chapter we analyze the pros and the cons of such applications. This will help in determining whether multimedia tours will gradually replace traditional audio tours as the mobile interpretation of choice in museums, or whether, due to the larger investment required in terms of infrastructure, maintenance, and content development, they will progressively diminish in use, leaving room for other solutions such as mobile phones or downloadable tours that, despite so far offering a more limited range of content, do not require the provision of additional hardware.

WHY MULTIMEDIA TOURS IN MUSEUMS?

The number of museums considering moving to multimedia tours is growing. This is mainly due to the potential that such solutions offer to museums and their visitors. Advantages include variety of interpretation, engagement of visitors, outreach to new audiences, support for orientation, and flexibility with content distribution. In this section we analyze in detail the reasons museums adopt multimedia tours.

Facilitating the Learning Process

Museums are being challenged to find new ways to connect people, ideas, and objects. Multimedia tours offer an opportunity to provide greater access to both the intellectual and cultural resources of a museum.[2] Digital images of related objects in storage and from other collections, filmed interviews with curators and artists, and contextual material in the form of photographs, film footage, music, and textual passages from related material can all be made available to the visitor at the touch of a button. Moreover, tours can combine exhibit interpretation with up-to-date and contextualized information about events, services, and other activities in the museum.

One multimedia tour that offers a wide variety of content is at Tate Modern. Here visitors have access to, among other things, audio commentaries about works as well as videos of artists explaining their work or commenting on other people's creations. On the PDAs provided, visitors can also play games intended to allow better understanding of some of the mechanisms behind the creative process and the meaning of the artwork under consideration. Particularly useful in this respect are the "touch and listen" applications, which permit visitors to click on highlighted elements of an artwork to discover more about its significance or symbolism.

By providing an in-depth and varied level of interpretation, such multimedia tours can be of great support to the visitor's learning process. An important principle of didactic communication is the use of several different channels of sensory information at the same time; with multiple channels, a learner can retain about three times as much information than with just one channel.[3] Furthermore, multimedia applications can be designed to support different learning styles or methods, to empower the visitor in asking questions and widening understanding, as well as to stimulate interaction, exchange, and participation.

Interactive programs on handhelds also offer a viewing "hook" for a reluctant visitor, together with the ability to choose and control what she or he wishes to listen to or view. "This can bring about a fundamental change to the nature of the visit, encouraging people to spend more time with individual exhibits, resulting in a sustained level of engagement."[4]

Multimedia is a proven learning tool outside of the museum context,[5] and specifically in museums, in the context of mobile interpretation, preliminary data seems to confirm that visitors using multimedia tours have more extensive learning experiences, demonstrate a deeper level of understanding and critical thinking, make more connections to their own history and background, and engage in greater personal learning.[6]

Serving the Needs of New Audiences

Another important reason museums have been attracted by multimedia tours since the mid-1990s is that such tours not only serve different types of visitors through a wider choice of information and content types but also appeal to new audiences who normally might not be interested in more traditional forms of interpretation such as audio tours.

An example is visitors with hearing impairments. Various museums, including Tate Modern, Versailles, Reina Sofia, Great Blacks in Wax, and the Museum of Science in Boston, have filmed commentaries using sign language for some of the objects in their collections or temporary exhibitions and made them available with captions or subtitles to visitors, using screen-based portable devices. In internal evaluations at Tate Modern and the Museum of Science in Boston, deaf visitors have shown a high level of satisfaction with these applications, which allowed them to better understand and appreciate the exhibition at their own pace rather than having to follow an interpreter.[7] (Ellen Giusti describes the Museum of Science implementation in greater detail in chapter 6.)

Other visitors who can benefit from the introduction of multimedia tours in museums are the young, who are regularly exposed to a high level of audiovisual information (music, videos, television, Web pages, advertising, etc.) and interactivity (video games, mobile phones, online chat, etc.). Multimedia tours are therefore particularly suitable for school visits, as shown by many projects and pilots. The idea is usually to encourage young people to explore cultural artifacts in a more dynamic and participatory way, and construct personal narratives to share with their classmates, family, and friends. One of the first implementations of this direction has been DIGIT, which enables students visiting the Dulwich Picture Gallery in London to use handheld devices for individual learning programs.[8] Teachers can preview the gallery's paintings on the Internet and set up specific trails for a class or for individual students. Once at the museum, each student receives a PDA already loaded with the trail. The system is intended to pose thought-provoking questions, which students have to answer by observing the paintings and then answering questions on the device. Back at school, each completed trail can be reviewed. In addition, further work and final portfolios can be created.

Another system developed specifically for school groups is MyArtSpace, which allows students on a school visit to discover, "collect," and annotate information about viewed objects on mobile phones distributed before the visit. (This is discussed in further detail in chapter 9.) Similar applications have been developed at Tate Modern, Chawton House,[9] and the Royal Insitution,[10] all in the UK.

The Gallo-Roman Museum in Belgium has a collaborative trading game for school groups in which, by answering questions or playing short games, participants have the possibility of gaining or augmenting exchangeable goods. The game is designed so that each player is dependent on the actions of other players; this is particularly important considering that handhelds have often been accused of isolating visitors and preventing social interaction (more on this shortly). In this case, observations confirm that the game provoked collaboration and interaction among its players.[11]

Further data from this evaluation show not only that interactive content (such as games, quizzes, and opinion polls) is by far the most popular educational material with students but also that the multimedia guides were successful in increasing students' engagement with objects. The questions prompted the students to find objects and study them in detail, enabling greater retention of information and a sense of ownership.[12]

Supporting the "Virtuous Circle"

Multimedia tours are often used to create continuity between pre-, post-, and actual visit experiences, through tools like bookmarking and annotation. This process has been referred to as the "virtuous circle."[13]

The Exploratorium, for example, was one of the first museums to experiment with technologies that allow visitors to take home "personal souvenirs" of the exhibition "to encourage them to continue the exploration at home."[14] This example has been followed by other museums including Tate Modern (see figure 5.1), which introduced bookmarking for their permanent collection multimedia tour in 2005;[15] the Museum of Science in Boston, which allowed visitors to save information in the "*Star Wars: Where Science Meets Imagination*" exhibition tour;[16] and at the City of Science and Industry in Paris, for the "That's Canada" exhibition in 2004.[17]

When integrated into the visitor experience, bookmarking can be a powerful tool for supporting the learning process,[18] creating a stronger relationship between the institution and the visitor. Research indicates that repetition is a major mechanism for retaining memories over time,[19] so bookmarking can increase visitors' knowledge about a collection or exhibition as well as stimulate a positive response about the visit and a desire to learn more.[20]

While bookmarking is not unique to multimedia tours, being found also on some audio tours as well as kiosks, screen-based handheld devices are well suited for this purpose because they not only allow visitors to select objects in situ but also enable visitors to review and reorganize bookmarks during the visit.

Providing Content Personalization

Serving different types of visitors can also be achieved through personalization, in which information is differentiated according to the specific needs and characteristics of each visitor.[21] Various types of visitors may require or desire information packaged in slightly different ways, with varying levels of interpretation. Technology is ideal for this since it can store a large amount of information, which can then be accessed and manipulated in a variety of ways.

FIGURE 5.1
Screenshot from the multimedia guide of the Tate Modern permanent collection. Copyright Tate, 2005 version

Examples   include   Hyperaudio,[22]   HIPS,[23]   Museum   Wearable,[24] ARCHEOGUIDE,[25] GUIDE,[26] and PEACH.[27] These prototypes have provided visitors with contextual and personalized information, thanks to the creation of "user models." A user model is a representation of visitors' behavior based on their interaction with the multimedia tour and their navigation in the museum. This model is exploited during the process of audio or text content delivery to describe objects or promote paths accordingly.[28]

For example, visitors to the Museo Civico in Siena, Italy, where the HIPS prototype was tested, could find items of interest, hear descriptions with references to items seen earlier, ask for additional information, and receive suggestions on alternative routes. The system generated audio messages dynamically, based on user requests, the history of browsing, the physical location at the moment of the query, and a set of preferences and information provided by the visitor at the beginning of the visit.[29]

Despite some positive feedback,[30] the complexity of these applications, which often require an efficient wireless network infrastructure as well as complex user-modeling algorithms, is such that most of these multimedia tour guides have been developed only as prototypes. This means that real personalization of multimedia guides is, as of this writing, at the conceptual stage and is likely to take some time before it can be practically introduced in museums.

Supporting Visitor Orientation

Finally, multimedia tours can support visitors' physical orientation and guidance in the galleries. This can be provided through static or interactive maps of the building that visitors access on screen-based devices, as well as through the use of wireless technologies such as infrared, RFID, Bluetooth, GPS, and WiFi. By automatically determining the position of the visitor as well as highlighting the objects nearby, these technologies, integrated with legible maps, can help visitors orient themselves and decide where to go next. They can also deliver content automatically based on location. Location-based services are particularly valued by large and complex museums with major visitor orientation problems, which is one of the main reasons museums consider multimedia in the first place.

Location-based services also afford greater possibilities for interactive content design. For example, bookmarking can be implemented using a wireless network, RFID, barcode scanning, or Bluetooth. In some projects, wireless connectivity has been used to provide interactive content (such as voting and polling), short messaging services (SMS) among visitors,[31] and content personalization. Location technologies can also be used for device tracking (e.g., if many people are clustered in one location, a message can be sent to these visitors suggesting other areas of interest); security alarms; and queue monitoring.[32]

ISSUES AND DIFFICULTIES

Despite the potential of multimedia tours, the number of museums that have effectively tested or introduced them so far is still relatively low, especially considering that more than a decade has passed since the first early experiments. This and the demise of major projects such as GettyGuide are a clear indication of the presence of obstacles and challenges.

Costs Issues

Development costs for multimedia tours are still high. While some costs have come down, it is still relatively expensive to create such tours, especially compared with traditional audio tours. Costs for hardware are typically higher for off-the-shelf touchscreen or keypad-based devices, smart phones, and iPods compared to standard, and even customized, MP3 audio players. Specific software is also necessary to display multimedia content, while extra software development is needed for bookmarking, polling, texting, context awareness, and so on. Moreover, to progress from pilot to actual installation requires long-life batteries, charging systems, and carrying cases.

Content development is definitely the most expensive aspect of creating a multimedia tour. Audiovisual resources and interactive material are far more expensive and time consuming than simple audio content, requiring programmers, designers, interface engineers, usability experts, and so on. Testing is required to ensure that the interface design and navigation are easy to use and understand by visitors. Costs for copyright clearance are also considerably higher for multimedia, especially when it includes images and videos of modern and contemporary pieces.

Hardware Issues

Touchscreen handheld devices are more delicate than other MP3 players, even when ruggedized. This requires more maintenance and often results in a higher turnover of hardware. In addition, because PDA processors generally have limited capabilities, especially when it comes to playing large content files (e.g., videos), there can be latency or freezing problems that affect the visitor experience. Another issue is that most multimedia tours have used off-the-shelf players for content delivery; this creates problems when a model is discontinued or updated. The operating system or screen resolution might change, meaning that content might not fit a new device, thus giving little continuity to projects over the longer term.

Next, there is the problem of upkeep. Often, museums engage in a one-time effort to develop an electronic guide. These systems grow out of date quickly as devices and exhibitions evolve. For example, Tate Modern had to plan its migration from one hardware manufacturer to another. The new devices provided better performance, but the screen resolution was different, meaning that content had to be adapted and new mass battery charging solutions had to be developed.

Multimedia tours also present more challenges in distribution. Staff members take longer not only to prepare devices for distribution but also to explain to visitors how to use them. This creates problems when turnover is high because it hinders the natural flow of visitors. Multimedia tours also require highly trained and dedicated staff members who know how to not only promote a tour but also address hardware problems quickly and efficiently.

Usability and User Perception Issues

Some of the most challenging issues with multimedia tours have to do with user perceptions and usability, which have deeply affected take-up rates.

The notion of "technology fatigue" is central. This is the experience of being exposed to so many new technologies so quickly that one is unable to keep up. Thus, many people are turned off by multimedia tours. Moreover, handheld devices often suffer from being too heavy and cumbersome to carry around. Visitors run into problems holding a device, and sometimes a stylus, while handling other things such as cameras, maps, or children, not to mention interacting with installations. This is particularly true in science museums, which are designed to be highly interactive.

It is not only the use of a more technologically advanced device per se that worries visitors. Most of the difficulties visitors encounter are at the interface level. Navigating complex hierarchies of content on small devices can lead to usability problems, especially if systems are not designed well, involving a cross section of real users in the development process. For example, the use of icons is problematic if people are not aware of their meaning. Also, in most cases a tour starts with a "home page" displaying menu options. Visitors not familiar with the idioms of the Internet (e.g., some older visitors) might not understand the notion of a home page and therefore might find it difficult to use the system from the start.

Multimedia solutions have so far failed to meet their considerable potential for supporting orientation and wayfinding. Maps are not necessarily any easier to use when presented on a small screen than in traditional printed form, particularly if their design renders information and icons too small to be read easily. For example, Metfinder was a research project at the Metropolitan Museum of Art in 2005, using a set of art objects in its American Wing. It tested the viability of providing audiovisual information as well as recommendations of other objects of interest. When a recommended object was chosen, the visitor could follow written, oral, or map-based directions to locate it. Visitors reported difficulty with all of these modes, but particularly with the maps. The reasons included the small size, the difficulty of interpreting symbols and signs, and the impossibility of telling exactly how to orient the map (how it should be held with respect to the direction the visitor is facing).[33]

Location technologies might help in this regard, making it somewhat easier for people to locate their current position and gain a sense of direction, but so far these solutions have not proven reliable or affordable. Many projects using location-aware technology have shown that the signal sensitivity is not specific enough to be useful and that the equipment requires too much maintenance. One of the most difficult issues is dropped signals as visitors navigate the museum. Findings suggest that visitors are intolerant of this, and further progress is needed to fulfill the promises of location technologies—especially to compensate for the higher investment and maintenance costs implicit in the use of these solutions.[34]

Learning and Social Issues

Despite some early evidence that multimedia tours are fairly successful in engaging the visitors and supporting personal learning, there are still many doubts in the museum community with regard to their effectiveness as an interpretation tool. One of the main concerns of museum professionals is that of "cognitive load."[35] While visitors can benefit from the varied and extensive content provided by multimedia tours, they can also quickly become overwhelmed with too much information that does not correspond with their interests or goals. Personalization techniques can help, but these systems are far from effective in implementation, meaning that cognitive load is still a significant issue.

Another issue of concern is that multimedia tours center around the "lure of the screen," the fear that screen-based devices distract visitors from looking at the exhibits.[36] There are contrasting views on this issue. Some professionals claim that visitors tend to look down at the computer screen, rather than up at the exhibits, when using multimedia tours.[37] Others provide evidence that rather than spending their time absorbed in the technology, visitors can positively engage with the objects and exhibits.[38] Ultimately, it depends on how the content is built: if visual material is introduced only when appropriate and without overwhelming the visitor, it can help the visitor "seek out details in the works, move back and forth between works, move closer to works, or point out details to family members and friends."[39]

Another major concern is that handhelds tend to lead to isolated, individualized experiences. As a response to this criticism, there have been some attempts to introduce text-based communication, the ability to see what other people have viewed or bookmarked, or games that require social interaction.[40] However, more research is needed to better understand whether collaboration is really desired by visitors and how to create collaborative learning content and activities that facilitate effective social interaction.

Finally, there is little evidence that multimedia tours so far support the "virtuous circle." Despite a few exceptions, the number of visitors using bookmarking and annotation features both during and after the visit is still limited. It is mostly confined to specialized target groups such as students and teachers who use the facilities in the context of a school visit.[41]

MULTIMEDIA TOURS: DEAD END OR WAY FORWARD?

Despite their great potential, the challenges of creating successful multimedia tours are numerous. While it is hard to predict how technology will evolve and how museums and audio tour companies will react, there are already indications that some of the problems can be addressed.

First of all, technology is getting cheaper, faster, and lighter. Hardware specifically designed for museums and nonexpert users is being developed by audio tour companies. This will solve some ergonomics issues, reduce costs, and provide more continuity of projects. The new devices, specifically built for museums, will guarantee more control over the operating system, screen resolution, and accessories. However, an issue is how reliable such custom solutions will be in practice. Do audio tour companies have the competence to produce such solutions cost-effectively?

Cost cuts can also occur at the software level. New software (content management systems or content assembly tools) that allow museums to assemble and update tours using existing content and a standard interface have been developed by all the major audio tour companies.

As far as learning, usability, and user perception are concerned, further research is required. Museums and audio tour companies are increasingly aware of the need to involve users in the development process, and this has already led to simpler systems and better interfaces. Similarly, the issue of technology fatigue will progressively decrease with time, as visitors become familiar with similar technologies in their everyday lives.

Clearly, challenges remain, and will for some time. These include, among other things, content development costs, maintenance, reliability of location technologies, and funding. Is the funding approach for these solutions right? Will museums understand that it is not a one-off commitment but that it requires regular investment and promotion? And is the museum flexible enough to react quickly to technological change?

While some of these problems will persist, others can be minimized by taking a more gradual approach to the development of multimedia tours. It is best to start with very simple solutions (locally stored, with manual content access, basic interfaces, and so forth), to test them adequately with the public, and then progressively add new features and services as confidence and expertise is gained. This allows museums as well as the public to gradually become used to the tools and the challenges they present.

One museum that has taken such an approach is Tate Modern, which went through a three-phase pilot introduction starting in 2002, before making the tour commercially available in 2005. The final result was a much-simplified version of what was originally tested, excluding location technologies, bookmarking, and text messaging, at least for the launch. The quality of the content and simplicity of the interface and navigation is a key factor in the success of this application.

Some institutions have tried to do too much with their multimedia guides, overcomplicating them by bundling too many specific features and services. As with any new initiative, testing and time are required to reach a successful formula. Future multimedia projects in large museums will be fundamental in determining the success of these solutions: if such museums take a gradual approach, devoting the necessary time to study the visitor experience, and focusing on content development rather than on the technology, they could cement the reputation of multimedia tours in museums and influence their future affirmation. If not, other museums may be discouraged from introducing multimedia tours for fear of repeating the same mistakes, thus seriously compromising the further integration of these solutions.

CONCLUSION

Despite the many obstacles that museums still face in developing multimedia tours, recent market developments and research show that these solutions have a promising future. However, their role as most museums' mobile interpretation device of choice will depend on a series of factors, including the reliability of custom-made solutions developed by specialized companies, the reliability of location technologies, the success of a few key projects that can serve as catalysts, the capacity for driving down costs, and a more flexible funding solution.

Another important variable that needs to be considered is that some museums are simply not suited for multimedia tours. For instance, not all museums put the same emphasis on interpretation. Some organizations have a high object turnover, which would require constant content development and updating. Others may feel that their public, being older and more traditional, may not be the right audience for an advanced museum guide. Even if this notion has been challenged by evidence from multimedia tour evaluations,[42] there remains a fear that older visitors might find it difficult to use multimedia guides.

Considering all these variables, it is safe to assume that, for the next several years, multimedia tours will probably continue to coexist with other solutions such as traditional audio guides, downloadable online tours, and mobile phone tours. In the long term, services offered on visitors' personal devices, including audio, text, and multimedia content, will become more prominent, allowing museums to save on distribution and hardware costs. However, this might take time for a series of reasons including quality of the experience, compatibility issues, and costs.[43]

Internationally, the business model for cell phone tours remains a challenge and will remain a challenge as long as the value chain is mediated by mobile network providers. In addition, museums cannot assume that every visitor will own or want to use a cell phone or any other personal digital device, so we are unlikely to see an immediate end to other platforms that are already familiar to museum visitors. The wider the variety of interpretation tools on offer, the more likely the museum is to reach a broad range of visitors.

NOTES

1. For a more comprehensive list of multimedia tours and pilots developed, see Nancy Proctor, "Off Base or on Target? Pros and Cons of Wireless and Location-Aware Applications in the Museums" (paper presented at ICHIM, Paris, 2005).

2. A. Manning and G. Sims, "The Blanton iTour—An Interactive Handheld Museum Guide Experiment," in *Museums and the Web 2004: Proceedings*, ed. David Bearman and Jennifer Trant (Toronto: Archives and Museum Informatics, 2004), at www.archimuse.com/mw2004/papers/manning/manning.html (accessed June 26, 2008).

3. Richard E. Mayer, *Multimedia Learning* (Cambridge, UK: Cambridge University Press, 2001).

4. Margaret Greeves, "Help at Hand: Working with Handheld Guides" (paper presented at Help at Hand: Working with Handheld Guides Conference, London, June 2006).

5. Mayer, *Multimedia Learning*.

6. Manning and Sims, "The Blanton iTour."

7. Nancy Proctor, "Access in Hand: Providing Deaf and Hard-of-Hearing Visitors with On-Demand, Independent Access to Museum Information and Interpretation through Handheld Computers" (paper presented at ICHIM, Berlin, 2004), at www.ichim.org/ichim04/contenu/PDF/4324_Proctor.pdf; E. Chin, "What Have We Learned from the *Star Wars* Multimedia Tour?" (paper presented at the ASTC 2006 Annual Conference, Louisville, Ky., October 2006) (accessed June 26, 2008).

8. Dulwich Picture Gallery, "E-Learning, Digit: A Technical Revolution at Dulwich Picture Gallery," at www.dulwichpicturegallery.org.uk/sackler/elearning.aspx (accessed June 26, 2008).

9. University of Southampton, "Creating the 'Outdoor Classroom,'" July 14, 2005, at www.soton.ac.uk/mediacentre/news/2005/jul/05_137.shtml (accessed March 26, 2008).

10. Chris Tellis, "Multimedia Handhelds: One Device, Many Audiences," in *Museums and the Web 2004: Proceedings*, ed. David Bearman and Jennifer Trant (Toronto: Archives and Museum Informatics, 2004), at www.archimuse.com/mw2004/ papers/tellis/tellis.html (accessed June 26, 2008).

11. Heleen Van Loon, Kris Gabriël, Kris Luyten, Daniel Teunkens, Karel Robert, Karin Coninx, and Elke Manshoven, "Supporting Social Interaction: A Collaborative Trading Game on a PDA," in *Museums and the Web 2007: Proceedings*, ed. Jennifer Trant and David Bearman (Toronto: Archives and Museum Informatics, 2007), at www.archimuse.com/mw2007/papers/vanLoon/vanLoon.html (accessed August 13, 2007).

12. Van Loon et al., "Supporting Social Interaction."

13. Alisa Barry, "Creating a Virtuous Circle between a Museum's On-line and Physical Spaces," in *Museums and the Web 2006: Proceedings*, ed. Jennifer Trant and David Bearman (Toronto: Archives and Museum Informatics, 2006), at www.archimuse.com/mw2006/papers/barry/barry.html (accessed June 26, 2008).

14. Exploratorium, *Electronic Guidebook Forum Report* (San Francisco: Author, January 13–14, 2005).

15. Silvia Filippini-Fantoni and Jonathan Bowen, "Bookmarking in Museums: Extending the Museum Experience beyond the Visit?" in *Museums and the Web 2007: Proceedings*, ed. Jennifer Trant and David Bearman (Toronto: Archives and Museum Informatics, 2007), at www.archimuse.com/mw2007/papers/filippini-fantoni/ filippini-fantoni.html (accessed August 13, 2007).

16. S. Hyde-Moyer, "The PDA Tour: We Did It; So Can You," in *Museums and the Web 2006: Proceedings*, ed. David Bearman and Jennifer Trant (Toronto: Archives and Museum Informatics, 2006), at www.archimuse.com/mw2006/papers/hyde-moyer/ hyde-moyer.html (accessed June 26, 2008).

17. R. Topalian, "Cultural Visit Memory: The Visite+ System Personalization and Cultural Visit Tracking Site," in *Museums and the Web 2005: Proceedings*, ed. David

Bearman and Jennifer Trant (Toronto: Archives and Museum Informatics, 2005), at www.archimuse.com/mw2005/papers/topalian/topalian.html (accessed June 26, 2008).

18. Mihaly Csikszentmihalyi and K. Hermanson, "Intrinsic Motivation in Museums: Why Does One Want to Learn?" in *Public Institutions for Personal Learning*, ed. John H. Falk and Lynn D. Dierking (Washington, D.C.: American Association of Museums, 1995), 67–78.

19. R. Brown and J. Kulick, "Flashbulb Memories," *Cognition* 5 (1997): 73–79.

20. Filippini-Fantoni and Bowen, "Bookmarking in Museums."

21. Jonathan P. Bowen and Silvia Filippini-Fantoni, "Personalization and the Web from a Museum Perspective," in *Museums and the Web 2004: Selected Papers from an International Conference*, ed. David Bearman and Jennifer Trant (Toronto: Archives and Museum Informatics, 2004), 63–78.

22. D. Petrelli, E. Not, M. Sarini, C. Strapparava, O. Stock, and M. Zancanaro, "HyperAudio: Location-Awareness + Adaptivity," in *ACM SIGCHI '99 Extended Abstracts*, Pittsburgh, May 1999, 21–22, at http://citeseer.ist.psu.edu/petrelli99 hyperaudio.html (accessed June 26, 2008).

23. P. Marti et al., "HIPS: Hyper-Interaction within Physical Space," in *Proceedings of the IEEE International Conference on Multimedia Computing and Systems*, vol. 2 (Washington, D.C.: IEEE Computer Society, 1999).

24. ICHIM, "Exploration of Some of Gauguin's Artworks by the Museum Wearable Adapted from the Digital Artistic Reproductions (DAR) Conceived by the Artist Etienne Trouvers," Paris, September 21–21, 2005, at www.ichim.org/ichim05/jahia/ Jahia/pid/647.html (accessed June 26, 2008).

25. Intracom, "ARCHEOGUIDE: Augmented Reality-Based Cultural Heritage On-site Guide," at http://archeoguide.intranet.gr (accessed June 26, 2008).

26. Lancaster University, "The Guide Project," at www.guide.lancs.ac.uk (accessed June 26, 2008).

27. Fondo Unico, "PEACH: Personal Experience with Active Cultural Heritage," at http://peach.itc.it (accessed June 26, 2008).

28. Silvia Filippini-Fantoni, "Museums with a Personal Touch," in *EVA 2003 London: Conference Proceedings*, ed. J. Hemsley et al. (London, 2003), 25.1–25.10.

29. Marti et al., "HIPS."

30. P. Marti, *The User Evaluation*, restricted report (Siena, Italy: Università degli Studi di Siena, 2000).

31. Tellis, "Multimedia Handhelds."

32. Tellis, "Multimedia Handhelds."

33. Metfinder, "Metfinder: A Handheld Solution for Independent Exploration and Discovery in the Museum" (paper presented at MCN Conference, Pasadena, Calif., November 2006), at http://72.5.117.137/conference/mcn2006/SessionPapers/Metfinder.pdf (accessed June 26, 2008).

34. Proctor, "Off Base or on Target?"

35. John Sweller, J. J. G. van Merriënboere, and F. G. W. C. Paas, "Cognitive Architecture and Instructional Design," *Education Psychology Review* 10 (1998): 251–96.

36. Proctor, "Off Base or on Target?"

37. Marjorie Schwarzer, "Art and Gadgetry: The Future of the Museum Visit," *Museum News* 68 (July/August 2001): 36–41.

38. Manning and Sims, "The Blanton iTour"; Kevin Walker, "Visitor-Constructed Personalized Learning Trails," in *Museums and the Web 2007: Proceedings*, ed. Jennifer Trant and David Bearman (Toronto: Archives and Museum Informatics, 2007), at www.archimuse.com/mw2007/papers/walker/walker.html (accessed June 26, 2008).

39. Manning and Sims, "The Blanton iTour."

40. Y. Laurillau and F. Paternò, "Supporting Museum Co-visits Using Mobile Devices," in *Proceedings of MobileHCI 04* (Berlin: Lecture Notes in Computer Science), 451–55, at http://giove.cnuce.cnr.it/pdawebsite/CiceroPublications.html (accessed November 2005); J. S. Cabrera, H. M. Frutos, A. G. Stoica, N. Avouris, Y. Dimitriadis, G. Fiotakis, and K. D. Liveri, "Mystery in the Museum: Collaborative Learning Activities Using Handheld Devices," in *Proceedings of MobileHCI 05* (New York: ACM Press, 2005), 315–18; Van Loon et al., "Supporting Social Interaction."

41. Filippini-Fantoni and Bowen, "Bookmarking in Museums."

42. Susie Fisher, "An Evaluation of Learning on the Move and Science Navigator: Using PDAs in Museum, Heritage and Science Centre Settings" (Bristol, UK: Nesta Report, 2005).

43. Silvia Filippini-Fantoni and Nancy Proctor, "Evaluating the Use of Mobile Phones for an Exhibition Tour at the Tate Modern: Dead End or the Way Forward?" in *EVA London 2007: Conference Proceedings*, ed. Jonathan P. Bowen et al. (London, 2007), 8.1–8.11.

6

Improving Visitor Access

ELLEN GIUSTI

The label "disabled" has become obsolete. People with disabilities are all around us. In fact, they *are* us. "Because current population estimates indicate that more than fifty million citizens [in the United States] have disabilities it is likely that persons with disabilities, as a significant segment of the general population," are potential if not actual museum visitors "whether or not they are explicitly identified as disabled."[1] Furthermore, "this category . . . can be entered by anyone at any time as the result of illness or injury and will be penetrated, inevitably, by anyone who lives long enough."[2]

In order to attract and hold audiences, museums must provide resources and technologies that acknowledge various cultures and abilities. Museums cannot operate under the old paternalistic model, a paradigm that implies they know what's best for their visitors.

It is well documented that museum visitors want a personally meaningful, relevant experience in which they feel in control. Building on this premise, this chapter uses the results of on-site visitor evaluations to demonstrate the potential of handheld technology to make museum experiences of mainstream, as well as nontraditional, audiences more accessible.

INTELLECTUAL ACCESS

Accessibility means more than physical access:

> While the publication of [disability regulations] . . . have led to significant
> changes in the industry, they predominantly focused on providing physical ac-
> cess to museums and did not address providing intellectual access to learning.
> Understanding physical differences among individuals and the resulting space
> and architectural requirements are important first steps. However, this infor-
> mation is not sufficient for providing true access to learning for all. Universal
> design for learning goes beyond physical accessibility. It involves creating mul-
> tisensory, multimodal learning experiences from which all visitors can learn by
> touching, seeing, listening, smelling, and sometimes even tasting.[3]

Since their inception, handheld technologies have been about accessibility
and its corollary, expanding audiences. Accessibility has many dimensions.
The word typically conjures images of people with disabilities—wheelchair
users and people with visual impairments—but in its broader meaning, accessibility is intellectual access for all. Intellectual access is a significant outcome for handheld audio interpretation.

When the first generation of handhelds was introduced in art museums in
1952, museums were perceived as elitist, the province of the cognoscenti.
Works of art were assumed to speak for themselves and needed no interpretation; regular visitors had to know about art and were to appreciate what curators displayed. Only tombstone-like information was provided: artist's
name, date, title (if available), and country of origin.

In the 1960s and 1970s things began to change. For one, the newly rich
wanted to participate in the cultural life of cities. And museums, pushed to
demonstrate their relevance and generate revenues, sought to attract more
paying customers. But these new audiences lacked the aesthetic upbringing or
background to interpret the information before them; this obliged museums
to reconsider their visitor provisions. Information was no longer sufficient;
user-friendly interpretation was required.

The 1980s boom in taped audiocassette tours is symptomatic of this shift.
The mellifluous voice of Philippe de Montebello welcomed the hoi polloi to
the Metropolitan Museum of Art in a tour that introduced visitors to curators
and curatorial decision making. And they learned about what was considered

important in art, and why and how artists influenced one another. There was no need to read scholarly books or journals. Visitors even learned the meaning of specialized terms like *chiaroscuro* and *trompe l'oeil*. But to follow the cassette tour, visitors were obliged to surrender a level of personal control over their visit and follow a predefined, linear path through the galleries.

Direct-access audio guides changed this. The ability to choose what to see, in what order, and then control the associated information flow is critical to a free-choice museum experience, as John Falk and Lynn Dierking describe in chapter 2. Direct access to information provided this, and this is now a basic expectation of all subsequent generations of handheld technologies.

Today, the desire for choice and control of interpretation is turning away from the curatorial voice to a more irreverent, dialogic approach in which everyone's opinion is valid—as indeed Peter Samis puts forth in chapter 1. If art is to be accessible to everyone, the reasoning goes, then everyone's interpretation is valid. Young people are not interested in the boring drone that has tended to dominate audio tours. They want to hear their peers, or some less institutional voice, and they want an engaging story to go with the art, something that makes it memorable by virtue of its humanity, not its historicity.

Consumer handheld technologies—technologies with which a visitor will already have a personal connection and control—have entered this field. No need to rent and familiarize yourself with an unknown technology; just bring your preloaded MP3 player or use your mobile phone to access new layers of information and interpretation.

ACCESSING CHILDREN

The pinball-like behavior of children in science centers is well observed. Rushing from one hands-on activity to the next, they appear to have little or no interest in understanding the underlying phenomena involved. To address this issue, and, in essence, improve intellectual access to the underlying phenomena, in 1997 the New York Hall of Science (NYHoS) designed a direct-access audio guide covering fourteen interactive activities. Its objective was to encourage engagement through a trial-and-error method of interaction. The assumption was that the audio could deliver underlying science concepts painlessly while the visitor manipulated the exhibit components.

The audio had a significant impact and success. Evaluations demonstrated that

• those with the audio guide spent over twice as long at an exhibit than those without, and their total visit length was some three times longer;
• visitors were better able to follow directions and understand explanations using the audio tour than using printed labels alone;
• audio tour users understood more than nonusers about the phenomena underlying hands-on exhibit components;
• audio tour users found the exhibition more fascinating than did nonusers; and
• most simply, children enjoyed the audio narration.

Museums are sites of informal education. Unlike schools, patrons are not obliged to learn. What visitors take away from a museum visit depends on their motivation and affective gains, and these, while difficult to measure, may be just as important as cognitive ones. The NYHoS evaluation concluded, "It seems clear that the current program for the tour . . . will make these 14 science experiences more readily available and intellectually accessible for people who select this modality to enhance their visit. 'I liked how my daughter stuck with it and figured it out,' said one father who used the audio tour with his child."[4]

PHYSICAL ACCESS

For certain audience groups, physical barriers pose insurmountable hurdles that impede them from reaching a context in which intellectual engagement with an exhibit can be attempted. While the size of this audience group is unknown, as highlighted above, this population potentially represents a significant portion of the museum audience and is one that is likely to grow.[5] In the United States, for example, the 2000 Census showed that close to fifty million people, or 19 percent of the population, reported that they had a "long-lasting condition or disability."[6] The Census also showed that the percentage of the population with a disability increases with age: while 19 percent of men and 16 percent of women aged sixteen to sixty-four reported having a disability, this rose to 40 percent of men and 43 percent of women for the population aged sixty-five and older.[7] By the year 2030, it is predicted, 20 percent of the

U.S. population will be over sixty-five years of age, as compared to 13 percent in the 2000 Census.[8]

Museums must increasingly consider the physical access requirements of this hitherto underserved audience group. Disability activists in the United States worked hard and long to achieve a major victory: the Americans with Disabilities Act of 1990 (ADA), which requires all places receiving public funding to be accessible to people who use wheelchairs, the blind, and the deaf, among other groups. It includes regulations for furniture construction so a wheelchair user can access displays, and videos and films must be captioned for the hearing impaired.

Access for the Blind

Millions of potential museum visitors live with visual disabilities that can interfere with their full participation in the cultural activities currently offered by museums. People over age sixty-five typically experience decreased visual acuity even when they consider themselves "normal" as opposed to disabled. How many times have we heard visitors complain that they cannot read the labels because of small print, low contrast, low light, and inconveniently low placement? These impediments affect all visitors but have special importance for aging visitors. With more disposable income and more leisure time, they are already a huge audience for museums. Conventional wisdom says that senior citizens are technologically challenged, or at least fearful of technology. This may no longer be true, now that seniors are working longer, using the Internet to access cultural and current events, and learning that the best way to stay in touch with grandchildren is via e-mail and cell phones. Yet, until recently, little has been done to provide access to museums for people who are blind or have low vision.[9]

Around 2000, the NYHoS developed a program designed to extend access toward the blind and visually impaired community. An important first step was to acknowledge that an audio tour created for the sighted audience cannot simply be marketed to blind visitors. It requires adaptation to provide greater audio descriptions of the exhibits to make them accessible physically as well as intellectually to this visitor group.

Visitors who are blind, or have low vision, have the same desire for personal control and free-choice learning as those who are sighted: when asked whether they preferred touring a museum with an audio guide or a sighted companion, the blind overwhelmingly chose the technology. They credited the audio guide with putting them in control of information flow—they did not have to "see" what their companion chose—and felt that audience-specific audio narration was more accurate than a fellow visitor reading labels.

It is easy to overlook that blind people will commonly state that they don't visit museums because there is nothing for them there to enjoy. Not without validity, the perception still exists that everything in a museum is behind glass or otherwise inaccessible to people who cannot see. This means that today there are virtually no visually impaired walk-in visitors. For them, without some sort of direct assistance, they cannot possibly have a successful experience.

In the NYHoS evaluation, one area where the blind wished they had greater control was wayfinding. The first iteration of audio interpretation for the blind at the NYHoS only gave visitors information once they reached an exhibit element; it did not tell how to find it. Wayfinding in museums is a challenge for *all* visitors but obviously more so for the blind. The majority of blind visitors who took the first iteration audio guide at NYHoS wanted more navigational help, suggesting, for example,

- "Describe how to get from point A to point B."
- "Add an audio map to the audio tour."
- "Provide directions to find facilities; for example, restrooms."
- "Add descriptions of the layout visitors would be walking through, reference points."

In response, incorporated into the second-iteration handheld guide was a user-activated beacon system, nicknamed "Ping!" Based on cell phone technology—a technology already used by most blind people—Ping! sought to enable visitors to navigate independently by following paths of "sonic breadcrumbs." When they reached their chosen destination, the system provided content and directions for using hands-on exhibit elements. The central aim was to deliver a system that allowed visitors to move independently among the hands-on exhibits and interact with them, based on the premise that people grasp science content best when they participate physically. Therefore, science museum visitors—including the visually impaired—must understand what they are supposed to do at exhibit stations for the experience to be effective.

Therefore, to further extend the physical access benefits the handheld guide could deliver, a talking tactile model of the NYHoS's full-scale Mercury/Atlas and Gemini/Titan rockets was created as a self-contained exhibit station and provided audio description about various parts of the rockets when touched. The rocket model could be activated by cell phone as well as touch, allowing a blind or low-vision visitor to launch the exhibit application without having to find the corresponding button that sighted visitors use. To achieve this, the exhibit was assigned a personal phone number; upon calling that number, the visitor was greeted by a computerized attendant who told them which key to press to begin. The visitor could then hang up the phone and interact directly with the exhibit.

The hardware for the second-iteration handheld guide was the visitor's own cell phone. Upon arrival visitors would turn on their phone and select a personal "Ping!" sound from among nine choices and then choose a destination from a menu of options including thirteen exhibits, the exhibit entry

FIGURE 6.1
Tactile rocket model at New York Hall of Science.

Steven Landau, Founder and Director of Research, Touch Graphics, Inc.

point, three facilities (cafeteria, coat check desk, and men's and women's rest-rooms), and the museum's admissions desk. Because a computer could track each participant's location, the menu changed accordingly, offering a choice of the seven nearest destinations at any time. Participants were instructed to ac-tivate their sound as frequently as needed. To reach more distant destinations, they used a sequence of audio beacons serving as "stepping stones."

Upon reaching a selected exhibit destination, a participant could press one of the numbers on the phone to hear prerecorded information about the ex-hibit, including how to use hands-on elements. The person could spend as much time as desired, visiting whatever number of possible destinations de-sired. They could stop at any point in time. The participant was then escorted to a meeting room for a posttrial interview.

Overall, the NYHoS system indicated that a handheld guide, incorporating a user-activated audio beacon system, could deliver wayfinding information and exhibit-related interpretation under naturalistic conditions (i.e., when several visually impaired users are simultaneously interacting with the system and when other visitors are in the museum). Participants' reaction to the sys-tem was very positive. In particular, the addition of "travel notes" (alerting users to such details as the number of steps on a route or indicating the di-rection of a staircase) was cited as a strength.

The two issues that created the most persistent problems were poor cover-age of all areas by cell phone service and "dead spots" in the network coverage that activates the beacons. The first problem varied depending upon cell phone service providers and cell phone models. When the network dropped out, the beacons ceased responding and the user was left stranded in the midst of the museum's typically noisy crowd of visitors. In this case, the participant had to request assistance.

Although a majority of visitors were pleased that the handheld guide was their own mobile phone, with which they were already familiar, the use of a cell phone in general was reported to be somewhat unwieldy. Specifically, holding the phone in one hand, and a cane or a dog's leash in the other, while navigating throughout the exhibit area, proved challenging at times. Other unpopular issues were the cost of calling from one's own phone and, for those who borrowed a phone, the lack of familiarity with the equipment.

Blind visitors who utilized the second-iteration handheld guide enjoyed the experience. They were enthusiastic about the potential to move through a

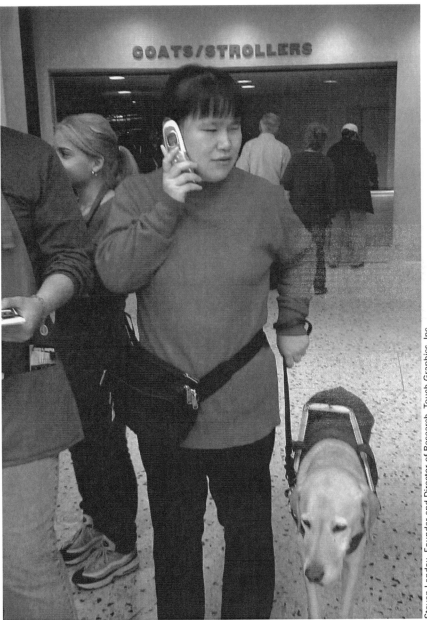

Steven Landau, Founder and Director of Research, Touch Graphics, Inc

FIGURE 6.2
Blind visitor with cell phone guide and guide dog.

public building on their own. However, though visitor response was positive, when assessing a technological innovation, one must be mindful of the fact that, like the mainstream population, people with visual impairments have varied interests and capabilities. Some of the participants were interested in hands-on science exhibits and some were not; some were technically adept and found the Ping! technology simple, while others, not comfortable with technology, found it challenging. Those who used the Ping! tour were quite diverse; they were similar only in that they were all visually impaired.

Access for the Deaf

Just as the Ping! system helps visually impaired visitors, signed or captioned tours can serve visitors who have hearing impairments. To this end, in 2006 Boston's Museum of Science decided to employ a handheld multimedia tour to accompany *Star Wars: Where Science Meets Imagination,* an exhibition about the real world meeting technologies featured in the *Star Wars* movies. A twenty-two-stop guided tour was produced in American Sign Language (ASL).

Like the Ping! system users, *Star Wars* tour users had diverse interests and varying learning modalities. And, similarly to mainstream *Star Wars* tour users, once inside the exhibition they used the tour in different ways. Some used it intensively and were absorbed in the content, while others barely used it at all. Some group members split up for the majority of the time inside the exhibition, while others stayed close together. But again, like Ping! users, the deaf participants said the ASL handheld tour was personally significant in many ways. "On one level, the ASL Tour represented museums' opening of their doors and embracing the deaf as valued visitors, and of becoming more inclusive places."[10]

Again, personal choice and level of control proved to be a central concern. The deaf participants said that the tour gave them a "sense of independence and control over what they saw, did, and learned."[11] Deaf visitors, like their blind and low-vision counterparts, appreciated not having to rely on companions to guide them whose agenda might not coincide with theirs. "The ASL Tour let me go at my own pace and explore topics of interest at greater depth like any other visitor," one said.

The museum found that people with disabilities perceive specialized handhelds differently from the general public. Rather than an add-on, for blind and deaf visitors the handhelds represent equal access to the exhibition's content. When asked how their experience without the handheld would be different, participants answered simply, "We wouldn't have any information." Because the ADA mandates equal access for all, deaf or blind visitors are not charged for tours, as are mainstream visitors. High museum charges already limit the access of many mainstream visitors to the content provided on handhelds.

Once they had a taste of an accessible tour, deaf and blind visitors wanted access extended to the rest of the museum. One of the deaf participants suggested offering a wayfinding mechanism to help visitors locate exhibits and show times. Participants thought it would be amazing if all museums had an ASL tour.

Ultimately, a sign-language tour can be seen simply as language translation, rather than a new program.[12] One could argue that for those who prefer to read, labels are a good source of information. However, research has shown that many members of the deaf community are not comfortable reading text and consider sign language their primary mode of communication. For this population, the choice is a false one because they have no other option but communicating in sign language. One caveat is that where audio and multimedia guides have been shown to lengthen the time visitors spend looking at artworks while listening to explanations, deaf visitors must spend that time instead looking at the handheld.

CONCLUSION

Handhelds have the potential to render museums more accessible. Over time, as museums become more welcoming to the general public, they are learning how to seek out and respond to their visitors' needs, both physical and intellectual, regardless of ability. All museum visitors want a personally meaningful, relevant experience over which they feel in control. Incorporating handheld digital technologies into museums has tended to individualize the visitor experience, providing an individual level of control over information flow, thereby rendering it personally relevant. Technology's future appears to have limitless prospects in terms of accessible experiences in museums.

NOTES

1. J. M. McNeil in C. J. Gill, "Invisible Ubiquity: The Surprising Relevance of Disability Issues in Evaluation," *American Journal of Evaluation* 20, no. 2 (1999): 2.

2. Gill, "Invisible Ubiquity," 2.

3. C. Reich and A. Lindgren-Streicher, "Universal Design Literature Review" (unpublished report for the Museum of Science, Boston, 2005).

4. Beverly Serrell and S. Tokar, "New York Hall of Science Random-Access Audio Tour Project" (unpublished evaluation report, 1997).

5. Reich and Lindgren-Streicher, "Universal Design Literature Review."

6. J. Waldrop and M. Stern, *Disability Status: 2000* (2003), at www.census.gov/prod/2003pubs/c2kbr-17.pdf (accessed April 11, 2007).

7. Waldrop and Stern, *Disability Status.*

8. Federal Interagency Forum on Aging-related Statistics (2000), in Reich and Lindgren-Streicher, "Universal Design Literature Review."

9. Reich and Lindgren-Streicher, "Universal Design Literature Review."

10. E. Chin and C. Reich, with contributions of A. Gaffney, "Lessons from the Museum of Science's First Multimedia Handheld Tour: The *Star Wars: Where Science Meets Imagination* Multimedia Tour" (unpublished evaluation report, 2006).

11. Chin and Reich, "Lessons from the Museum."

12. Chris Tellis, "Multimedia Handhelds: One Device, Many Audiences," *Museums and the Web 2004: Proceedings*, ed. David Bearman and Jennifer Trant (Toronto: Archives and Museum Informatics, 2004), at www.archimuse.com/mw2004/papers/tellis/tellis.html (accessed June 26, 2008).

7

Structuring Visitor Participation

KEVIN WALKER

Every museum visitor is a storyteller with authority. Every evocative object on exhibit is a mnemonic device. Every visitor interaction is story-making as visitors fit portions of our collections into personal frames of reference; most often in ways we neither intended nor anticipated.

— *Robert Archibald, Missouri Historical Society[1]*

This chapter examines ways in which personal digital technologies support active visitor construction, collaboration, and sharing. I pay special attention to one way of structuring visitor activity—personalized learning trails. Visitors use their own mobile devices to tag and annotate particular artifacts, exhibits, or places for later reflection, editing, and sharing, all within a simple trail structure. Structures such as trails can act as effective mental models for museum visitors, especially when

- they have a narrow subject focus and manageable amount of data capture;
- they are created as a narrative or a conversation; and
- emphasis is on construction, not mere data capture—in other words, making a trail for someone else to follow helps to build your own knowledge.

Tools change such activities, and in this case portable digital technologies make capturing, editing, and sharing easier through the systematic recording and representation trails.

In free-choice learning environments such as museums, as John Falk and Lynn Dierking describe in chapter 2, learning can be incidental: visitors do not always go to a museum with an explicit or specific goal in mind. But as Jeffrey Smith and Pablo Tinio describe in chapter 4, visitors desire a mix of structure and freedom, and this chapter shows that introducing a goal or task, even for casual visitors, can focus learning while also producing a product that can be shared or assessed thereafter.

SOCIAL ACTIVITY, PERSONAL LEARNING

The very word "visitor" connotes passivity—someone who visits a collection owned by a museum, then goes away—and unsurprisingly, this term is used more often by art museums than science museums. But as noted elsewhere in this book, many museums now accept the "constructivist" view that knowledge is actively produced by a learner, focusing not on what an individual learns but on what the museum contributes to his or her existing knowledge.

Museums have an inherent conflict with this view, however, because of their role as authoritative cultural institutions. To accord visitors with the power to construct their own meanings is to undermine curatorial authority. In so doing, a museum not only risks losing its authority, but it also absolves itself from interpretive responsibility for the meanings it produces and circulates in the culture.[2]

If the "constructivist museum" is taken to its logical conclusion, according to Juliette Fritsch, head of gallery interpretation, evaluation, and resources at London's Victoria and Albert Museum (V&A), "then surely there is no museum knowledge except for that which the visitor constructs in his or her head." She therefore advocates "interpretive communities" centered around particular artifacts, in which the museum retains the central—but not the only—voice.[3]

Constructivism alone is not enough. It assumes that meaning making and knowing resides inside a person's head, and thus tends to downplay not only social but also environmental factors—Falk and Dierking's other two contextual spheres. "The process of meaning making is the process of making sense of experience, of explaining or interpreting the world to ourselves and others. In museums, meaning is constructed from objects, and from the sites themselves."[4]

Museum objects are encountered in the context of other (usually similar) ones, along with interpretive information and tools, and so cannot be considered in isolation. And since museum visits generally last no longer than a few hours, "dwell time" at any single exhibit or object is usually less than thirty seconds.[5]

> A story is not so much about the artefact itself, rather about how it came to be here and what is its relationship to other objects. There are many stories to be told and different perspectives from which they can be told, and these stories often overlap with others. We have further come to understand that there is seldom a "true story," as curators describe parts of their research to be almost like "detective work." Thus information exists in several layers.[6]

Thus, learning, too, has a historical context. This is explained well by Mike Sharples:

> A visitor to an art gallery stands in front of a painting. She has arrived at a current understanding of the painting from the path she has taken through the gallery—taking in the ambience, stopping at other paintings, reading the guidebook—and also from a lifetime of creating and interpreting works of art starting with childhood drawings.[7]

Most people go to museums expecting to learn something. They usually do not have a predetermined idea of just what they are going to do or learn, unless they know a lot about the works; thus, they willingly allow the museum to structure their visit to some extent (see chapter 4). But different audiences use museums differently. Tourists, for example, tend to try to see the entire museum; schools, on the other hand, tend to target visits to a particular gallery or a subject. The self-directed lone visitor may have a specific goal related to long-term interests in, say, Rembrandt's portraits, and may be building on long-term knowledge. In most cases, visits are brief and leisure oriented. But according to Falk and Dierking, deep, transformative learning occurs over the long term; therefore, people need reinforcing experiences.[8]

MAKING CONNECTIONS

It was the connectionist aspect of the mind that prompted Vannevar Bush to envision, in 1945, a trail-based external memory system.[9] His idea for a "Memex" that served as a hypertext library on a researcher's desk helped to inspire the World Wide Web. And the Web now provides a link between museum visits and pre- and post-visit learning.

Museum collections, online databases, and the Web, generally, are all unstructured or semistructured information spaces with little to guide a user in navigating through them. Technology provides an advantage in Web navigation over place-based learning, with structured searches and a recorded "history" of links visited. But this digital support is not semantic—it offers no support with regard to the meaning in the links that Falk and Dierking and others mention. The technologies hold great potential but not without some structure around their use.

Most museum visitors are familiar with browsing the Web, and the use of Web-based conventions such as page-based navigation and hyperlinks has influenced visitors' comfort with personal digital assistant (PDA)-based guides at the Dulwich Picture Gallery[10] and Tate Modern (see chapters 3 and 5). However, portable technologies continue to be used primarily for one-way information delivery. My research, however, shows that giving visitors control of the technology helps to develop understanding through play, creation, critique, and collaboration.

This shift of focus from content delivery to social construction reflects a societal shift in digital media from centralized control to user-generated content and personalized learning. In an age when personal monologues, dialogues, and multilogues proliferate on the Internet, and museums are adapting accordingly, the social dimension of learning becomes paramount.

Technologies mediate visitors' meaning making. But as Ben Gammon and Alexandra Burch report in chapter 3, a frequent problem with mobile digital devices in museums is the mismatch between the content on the device and the visitor's real-world experience in the museum. This is why portable technologies in museums are often targeted at younger audiences—because these groups adopt new technologies more readily, and there is a perception that new technology can hook them in to learn about the old stuff (objects, artifacts, and paintings). Indeed, Dulwich Picture Gallery has found that teenage students initially are engaged primarily by the PDAs given to them, but these soon become almost second nature to them and serve to focus them on learning about specific aspects of art and art history.[11]

This mirrors the ease with which nine-year-olds assimilated mobile phones for data capture and interpretation at Kew Gardens (a large botanic garden in London), in research I conducted. The garden staff initially thought the technology would distract the students from "the real thing" (plants in this case),

FIGURE 7.1
Capturing photos and audio annotations focused students' learning about food plants.

a common fear expressed by museum educators. But everyone agreed afterward that the technology instead served to focus the students and keep them engaged in learning.[12]

An evaluation of PDA trials at four diverse UK museums found that "the PDA is at its strongest with young people in home/leisure mode. There may also be an opportunity to hook in non-visitors to museums, who like learning through entertainment and have become disenchanted with the traditional museum agenda." The potential for engagement, however, is not limited to the young. "We might have expected traditional older museum visitors to be the least receptive to PDA interpretation and they certainly have the most difficulty with the concept and the initial use. Experience at the Fitzwilliam [Museum in Cambridge], however, shows otherwise. Traditional visitors found interpretation through the PDA to be absorbing, enlightening and valuable."[13]

BOOKMARKING

Handheld technologies such as PDAs can add to the surfeit of information already in the museum environment. According to the PDA study, "People are as likely to be trying to contain the amount of information they have to take in, as much as they are trying to expand it."[14]

Bletchley Park addresses this by allowing visitors to use their own mobile phones to tag particular exhibits by sending text messages, for later retrieval of information on the Web. The museum decided that much of its collection (related to code-breaking during World War II) was best suited to post-visit reflection; its extensive archive includes many interviews. At the museum's website, a visitor enters his or her mobile number and gets a list of the exhibits bookmarked, along with thematic trails linking off of each. Linked topics are suggested based on statistical co-occurrence, and keywords are used to automatically find the paths between individual resources. But users can also construct their own pathways, and thus the system includes both top-down and bottom-up aspects linking curatorial and user-generated content.[15]

Visitors—especially teachers—consistently express a desire for the capability to tag objects or exhibits, for further exploration from home or school (see Sherry Hsi, chapter 8). However, only 15 to 20 percent of casual visitors typically visit the museum's website to retrieve data in the museum they have tagged for later, even with the use of personalized information such as photos or video of visitors, according to one study. "Research is therefore needed to understand what activities will be successful in providing the impetus for visitors to extend their journey into the virtual space. Museums therefore need to anticipate ways in which visitors will create their own connections and deliver the tools and services that will facilitate this."[16]

Centre Pompidou in Paris has confronted this issue by engaging an audience committed to serious criticism, instead of merely soliciting feedback or bookmarks from anyone. It created a tool that enables engaged visitors to collect and arrange individual interpretations into a narrative form. This was initially intended for critics and serious amateurs to create "signed viewings" of works, but it makes possible commentary and criticism of any artwork or exhibit.[17] Visitors can use any phone to record audio messages about particular exhibits. The effect is similar to what Ellen Giusti describes in chapter 6 about a guidance system for the blind, in which exhibits have their own phone number; in this case, a voice mail extension. After recording commentary about particular works, a visitor can later edit and arrange these into a "signed viewing" at the museum's website by entering his or her phone number.

THE VALUE OF THE VOICE

In my research, audio has proven to be particularly important. As a teacher pointed out, recording audio into a small device not only benefits learners with literacy problems but also is just generally an easier way for anyone to make notes while moving around than to write something down—or worse, to type it into a tiny keypad. It is also faster, mentally, to simply say something rather than write it; though this is a different cognitive process, as discussed shortly.

In my early learning trails research, audio recorders were given to museum visitors to record their thoughts at particular objects or exhibits of interest to them. Unfortunately, this resulted mostly in recordings of someone either reading an object label or saying it appealed to them. "It's nice" or "I like it" tended to be typical comments. In a few cases they might relate it to their own personal knowledge or experience—"This reminds me of . . ." for example.

This demonstrated a few things. One was that the label information was probably the most useful: it provides a means of location tracking without having to use sophisticated technology to do so; if the locations of particular objects are known, a visitor's physical trail can be traced from the label information. But this raised a further question: how important is the physical trail?

If there is a trail that links objects thematically or conceptually, the actual location of those objects might only be of interest if the visitor (or someone else) is going to follow the trail again. It had been assumed that a map plotting a visitor's path through a space would be an important representation, but in fact, more important is whether the trail is coherent in a thematic or narrative sense, not a spatial one.

These initial trails also showed that there was no incentive to include particular objects without some structure or theme to the trail; otherwise, it becomes merely a series of disconnected recordings. Hence, successive iterations of the research narrowed the focus each time, or imposed more structure, such as assigning roles or activities—for example, "Take a picture of the object," or "Interview someone else who is looking at the object."

Finally, from the initial trials it was discovered that novice visitors with little or no knowledge about a particular artifact or subject area need a place to start—such as a curatorial interpretation to react to. Thus, at particular locations, a museum can deliver some information about an object (as most audio and multimedia tours do) but can then solicit the visitor's reaction to that information.

Even better to have an actual curator standing next to the visitor for interrogation, for real learning happens in conversation with an expert, a more able learner, or failing that, a peer. At my first Kew Gardens trials, there were not enough mobile devices (in this case, phones) for all the students in a class. So they were put into groups of five, each with a phone running MyArtSpace software (which is described in detail in chapter 9).

The results were surprising and interesting. Groups used the phone in different ways—in some cases, a teacher or adult handled the device and asked questions of the students; in other cases, students took turns making recordings; and in others, one eager student appropriated the device and acted as reporter. In one activity, each group was assigned a food plant and had to make up a rhyme or riddle about it for the other groups to guess the plant. This resulted in recordings of whole groups singing, rapping, or reciting in unison into the phone. But the most interesting cases were when two students paired up, and one interviewed the other, or when they rehearsed and recorded a scientific observation.[18]

This is further evidence that, contrary to traditional museum fears, portable devices do not seem to mitigate against social interaction, particularly when they are shared. Co-located learners often prefer to gather around a shared screen than to collaborate using individual screens.[19] Mobile devices enable carrying findings physically to other learners to share—this is easier than, for example, sending a picture or a link to another co-located learner.[20] Putting multiple handheld screens next to each other for comparison allows learners to build joint understandings; handhelds can serve to bridge private and public interactions.[21] And Paul Rudman et al. (in chapter 9) found the mobile phone, with appropriately supporting software, to be a "conversational partner" that prompted discussion among students.

For research purposes, it can be useful to record a person's or group's entire visit. This is just what one researcher did at the Exploratorium, a hands-on science center in San Francisco. "Hearing or reading visitors' complete conversations is a vivid experience that brings one right into the arena where real museum learning occurs. . . . It is much easier to understand visitors' personal or social contexts when studying a half-hour of their conversation than the few minutes typical of a single-element interaction."[22] Conversations between visitors veered between knowledge, the situation, actions to be taken, and descriptive language. Most of the time, visitors (unsurprisingly) talked about the exhibit in front of them, and only 5 percent made any conceptual connections between exhibits. A bit of structure, such as a trail, can help visitors make those connections between exhibits, combining personal experience with the newly encountered into a more coherent story.

This kind of narrative construction takes place in dialogue. Visitors are "people who are in conversation, literally and figuratively, with the artwork on display and with the curatorial intent."[23] Interpretation is negotiated in the form of dialogue—a tacit dialogue between visitor and curator,[24] an internal dialogue,[25] a dialogue between the individual and environment,[26] or actual, external conversations between people collected around an artifact or connected by technology.[27] Some view a museum visit itself as one long conversation, which continues after visitors leave the museum.[28] Others look beyond the exit doors:

> In discussing learning in museums many of us have referred to the "conversations on the way home." We know what we mean by this (or think we do) but we are frustrated by not being able to capture the language and thoughts by which the unique and special experience is folded into, and becomes part of, the

visitor's larger sense of self—the manner in which the details become appropriated by the visitor and the results of that appropriation.[29]

Conversation changes before, during, and after a visit, reflecting visitors'
identity, knowledge, and engagement. It is both an instrument and indicator
of learning—a methodology for building knowledge. An individual's mind is
socially developed through interaction with others, and language is a negoti-
ating medium for teaching and learning, with meaning being made after, not
before, words are uttered. Thus, language is a tool—indeed the "tool of
tools."[30] A visitor's identity is therefore not fixed but rather defined in the museum through artifacts and the personal experiences and knowledge they
evoke.[31]

Crucially, however, this does not mean that learning is more effective in
shared, as opposed to solitary, encounters with museum objects. Rather,
meaning is constructed through human activity, which is always socially situated—often through discourse, either explicit or implicit, mediated by language. For example, museum educators often mediate visitors' encounters
with objects. In such situations, "it is the way in which discourse is guided and
supported that is most significant for meaning-making activity."[32]

Technology, of course, also serves as such a mediator, guide, and support,
and can also capture the "conversations on the way home" referred to earlier.[33]

TALKING BACK

One method of extending the visit is soliciting visitor feedback. The Science
Museum in London has been successful in getting visitors to not only contribute written feedback but also read the contributions of others, even spending long periods of time at feedback stations in the galleries. Parents on family
visits especially value this as a tangible way for their children to reflect upon
and articulate what they have learned.[34]

Many museums provide the means for feedback or visitor contributions on
their websites. However, take-up is typically low. For example, the V&A solic-
its website visitors' memories of their first visit to the physical museum.[35]
However, as of November 2007 this contained only six entries, at least half of
which were by museum staff. This means that either no one is contributing or
contributions are moderated and no one on the museum staff is moderating
and posting this content.

Thus, an in-gallery or online feedback forum is successful only to the extent that it is a true dialogue.

> A dialogue is possible not just when people begin to speak, but when they start to listen. Despite many museum efforts to encourage people to comment, speak up, and have their say, it is unclear whether anyone is actually listening. In many institutions, only the floor staff and the marketing department are actually interested in what visitors have to say.[36]

By contrast, the V&A's "Design your own Arts and Crafts tile" online activity unwittingly created an active online community, as people used the online gallery to not only post the designs they created but also communicate with each other, through the text descriptions of their designs (since the site was not intended for communication between visitors). The site became very popular, linking remotely located participants with shared tastes and interests, and spawning lasting friendships.[37] This is perhaps because the Web visitors were concentrated in a tightly focused community and because they were given tools for active construction and communication. They needed no dialogue with the museum staff, instead using tools the museum provided to create their own community.

NARROWING THE FOCUS

In the Kew Gardens trials, there was another unexpected constraint: it was discovered that, in the MyArtSpace software, the duration of a single audio recording was limited to fifteen seconds, in order to minimize file size and uploading time. This, of course, was a challenge—but it turned out to be the right kind of challenge because it meant that the learners had to really think about what they were going to say before recording it. So, for example, a pair of girls could be observed meticulously scripting and rehearsing a recording; a pair of boys arguing over exactly how to describe a particular plant; and a parent soliciting descriptions from each student in a group. Again, more structure (this time technology-imposed) resulted in more interesting, and more usable, results. While an entire recorded visit is valuable for researchers, no visitor is going to be motivated to podcast or even listen to such a recording, unless it is scripted or edited into digestible and meaningful chunks.

Tate Britain has an online activity called "Create your own collection." A subset of images from the museum's collection database is available, from which the visitor chooses six, then gives the collection a name and can send via e-mail or print as a two-sided leaflet that locates the chosen artworks on a floor plan. As inspiration, the museum provides samples such as the "I like yellow collection" and the "odd faces collection."

Theming a collection, and restricting it to six artworks, is an appropriate level of structure. Psychological research has shown that a person can only hold about seven items in short-term memory.[38] For this reason, in a later trial at Kew, between six and nine trail stops were specified, and this proved a manageable scope. Bletchley Park also suggests that visitors capture a similar number of bookmarked items. Tasks with a narrow scope and data collection requirements are particularly important for casual visitors, since museum visits are, as Falk and Dierking point out in chapter 2, leisure oriented and generally brief.

Similarly, any activity that requires going to a fixed computer to complete it may not get finished. In some of my first trials, learners were supposed to go back to the website after the visit, and never did. In a successive trial, time was built in at the end in the on-site computer lab. But this is really not practical for anyone but a class of students who can be told what to do. Therefore, if any trail editing or other post-visit tasks can be conducted directly on the mobile device, this is ideal because they can be done anytime, anywhere. Mobile technologies now facilitate these tasks, with large touchscreens and always-on connectivity.

Effective learning is not related to the amount of data we can pack into our heads. In museums particularly, the amount of information is not as important as the conceptual framework into which it is placed; indeed, such frameworks not only facilitate information integration but also constitute most of what is remembered.[39] This is what good museum exhibitions do—present individual objects within a structure that links them together to form a larger picture. Some exhibitions make this structure explicit with advance organizers and orientation guides, and some provide themed trails for school groups, families, or individual visitors, usually printed on paper. But my research has found that real learning can occur when visitors are actively engaged in constructing their own learning trails for reflection or sharing.

CONCLUSION

Mobile technologies have the potential to support visitors' meaning making by framing and focusing their activities (through structures such as trails) and interactions (with objects and other people). These same technologies enable easy and automatic linking to virtual communities outside the building and after the visit.

An open question is just how much structure is appropriate for school groups and for casual visitors. In keeping with Smith and Tinio's findings and Peter Samis' experience (both in this volume), it appears that, for both groups of visitors, more structure and narrower scope seem to contribute to greater learning as well as increased incentive to participate. In other words, museum visitors learn more, and are more inclined to contribute and share, when their activities are concentrated on specific subjects and on a limited number of objects or exhibits. And learning occurs when museums cease to view visitors as passive containers and begin recognizing them as active constructors—not only of meanings inside their heads but also of connections and creations in the world, on the screen, in the museum, and beyond.

NOTES

1. Robert R. Archibald, "Touching on the Past" (paper presented at the Social Affordances of Objects seminar, London, November 2006).

2. Cheryl Meszaros, "Now THAT Is Evidence: Tracking Down the Evil 'Whatever' Interpretation," *Visitor Studies Today* 9, no. 3 (Winter 2006): 10–15.

3. Juliette Fritsch, "Thinking about Bringing Web Communities into Galleries and How It Might Transform Perceptions of Learning in Museums" (paper presented at the Museum as Social Laboratory: Enhancing the Object to Facilitate Social Engagement and Inclusion in Museums and Galleries seminar, Arts and Humanities Research Council Seminar series, London, March 2007).

4. Eilean Hooper-Greenhill, ed., *The Educational Role of the Museum*, 2nd ed. (New York: Routledge, 1999), 12.

5. Valorie Beer, "Great Expectations: Do Museums Know What Visitors Are Doing?" *Curator* 30, no. 3 (1987): 206–15; C. Cone and K. Kendall, "Space, Time, and Family Interactions: Visitor Behavior at the Science Museum of Minnesota," *Curator* 21, no. 3 (1978): 245–58.

6. J. Halloran, E. Hornecker, G. Fitzpatrick, D. Millard and M. Weal, "The Chawton House experience: Augmenting the grounds of a historic manor house" (paper presented at Re-thinking Technology for Museums: Towards a New Understanding of People's Experience in Museums, Limerick, Ireland, June 2005).

7. Mike Sharples, "Learning as Conversation: Transforming Education in the Mobile Age" (paper presented at the Conference on Seeing, Understanding, Learning in the Mobile Age, Budapest, Hungary, April 2005).

8. John H. Falk and Lynn D. Dierking, *Learning from Museums: Visitor Experiences and the Making of Meaning* (Walnut Creek, Calif.: AltaMira, 2000), 131.

9. Vannevar Bush, "As We May Think," *Atlantic Monthly*, July 1945.

10. Ingrid Beazley, "Spectacular Success of Web Based, Interactive Learning," in *Museums and the Web 2007: Proceedings*, ed. Jennifer Trant and David Bearman (Toronto: Archives and Museum Informatics, 2007), at www.archimuse.com/mw2007/abstracts/prg_325001001.html (accessed November 5, 2007).

11. Beazley, "Spectacular Success."

12. Kevin Walker, "Visitor-Constructed Personalized Learning Trails," in *Museums and the Web 2007: Proceedings*, ed. Jennifer Trant and David Bearman (Toronto: Archives and Museum Informatics, 2007), at www.archimuse.com/mw2007/papers/walker/walker.html (accessed June 26, 2008).

13. Susie Fisher, "An Evaluation of Learning on the Move and Science Navigator: Using PDAs in Museum, Heritage and Science Centre Settings" (Bristol, UK: Nesta Report, 2005).

14. Fisher, "An Evaluation of Learning."

15. Bletchley Park, "Bletchley Park: National Codes Centre," at www.bletchleypark.org.uk; Paul Mulholland, T. Collins, and Z. Zdrahal, "Bletchley Park Text: Using Mobile and Semantic Web Technologies to Support the Post-visit Use of Online Museum Resources," *Journal of Interactive Media in Education* (December 2005), at www-jime.open.ac.uk/2005/21/ (accessed November 5, 2007).

16. Alisa Barry, "Creating a Virtuous Circle between a Museum's On-line and Physical Spaces," in *Museums and the Web 2006: Proceedings*, ed. Jennifer Trant and David Bearman (Toronto: Archives and Museum Informatics, 2006), at www.archimuse.com/mw2006/papers/barry/barry.html (accessed June 26, 2008).

17. Vincent Puig and Xavier Sirven, "*Lignes De Temps*: Involving Cinema Exhibition Visitors in Mobile and On-line Film Annotation" (paper presented at Museums and the Web 2007, San Francisco, April 2007), at www.archimuse.com/mw2007/papers/puig/puig.html (accessed November 5, 2007).

18. Walker, "Visitor-Constructed Personalized Learning Trails."

19. Dana Cho, "Grasping the User Experience" (paper presented at Collaborative Artifacts Interactive Furniture, Château-d'Oex, Switzerland, June 2005).

20. Pierre Dillenbourg, "Scripted Collaboration and Locative Media" (paper presented at the Swiss Summit on ICT: New Generation Networks, Fribourg, Switzerland, October 2005).

21. P. Vahey, Jeremy Roschelle, and D. Tatar, "Using Handhelds to Link Private Cognition and Public Interaction," *Educational Technology* (May–June 2007): 13–16.

22. Sue Allen, "Looking for Learning in Visitor Talk: A Methodological Exploration," in *Learning Conversations in Museums*, ed. Gaea Leinhardt, Kevin Crowley, and Karen Knutson (Mahwah, N.J.: Erlbaum, 2002), 259–303.

23. C. Stainton, "Voice and Images: Making Connections Between Identity and Art," in *Learning Conversations in Museums*, ed. Gaea Leinhardt, Kevin Crowley, and Karen Knutson (Mahwah, N.J.: Erlbaum, 2002), 214.

24. Stainton, "Voice and Images."

25. Hooper-Greenhill, *The Educational Role of the Museum.*

26. Falk and Dierking, *Learning from Museums*, 136.

27. E. Van Moer, "Talking about Contemporary Art: The Formation of Preconceptions during a Museum Visit," *International Journal of the Arts in Society* 1, no. 3 (2006): 1–8; Etienne Wenger, "Communities of Practice: Learning as a Social System," *Systems Thinker* (June 1998), at www.co-i-l.com/coil/knowledge-garden/cop/lss.shtml (accessed November 5, 2007).

28. Paulette McManus, "It's the Company You Keep: The Social Determination of Learning-Related Behavior in a Science Museum," *International Journal of Museum Management and Curatorship* 53 (1987): 43–50.

29. Gaea Leinhardt, C. Tittle, and Karen Knutson, "Talking to Oneself: Diary Studies of Museum Visits," in *Learning Conversations in Museums*, ed. Gaea Leinhardt, Kevin Crowley and Karen Knutson (London: Erlbaum, 2002), 104.

30. L. S. Vygotsky, "The Genetic Roots of Thought and Speech," in *Thought and Language*, ed. and trans. A. Kozulin (Cambridge, Mass.: MIT Press, 1986); L. S. Vygotsky, *Mind in Society: The Development of Higher Psychological Processes* (Cambridge, Mass.: Harvard University Press, 1978; published originally in Russian in 1930), at www.marxists.org/archive/vygotsky/works/mind (accessed November 5, 2007); M. Halliday, "Towards a Language-Based Theory of Learning," *Linguistics and Education* 5 (1993): 93–116; James V. Wertsch, *Vygotsky and the Social Formation of the Mind* (Cambridge, Mass.: Harvard University Press, 1985).

31. Falk and Dierking, *Learning from Museums*.

32. Palmyre Pierroux, "The Language of Contextualism and Essentialism in Museum Education" (paper presented at Re-thinking Technology in Museums: Towards a New Understanding of People's Experience in Museums, Limerick, Ireland, June 2005).

33. Leinhardt, Tittle, and Knutson, "Talking to Oneself."

34. Alexandra Burch and Benjamin M. Gammon, "The Museum as Social Space: Scaffolding the Scaffolder," 2006, at www.kcl.ac.uk/content/1/c6/02/36/86/burch.pdf (accessed November 5, 2007).

35. Victoria and Albert Museum, "History of the V&A: Your First Visit to the V&A," at www.vam.ac.uk/collections/periods_styles/history/first_visit/index.php (accessed March 27, 2008).

36. Andrea Bandelli, "Talking Together: Supporting Citizen Debates," in *Many Voices: The Multivocal Museum*, at http://clearingatkings.com/content/1/c6/02/37/03/Bandellipaper.pdf (accessed November 5, 2007).

37. Gail Durbin, "Is There Anyone Out There? Finding Out About How Our Web Sites Are Used," in *Museums and the Web 2006: Proceedings*, ed. Jennifer Trant and David Bearman (Toronto: Archives and Museum Informatics, 2006), at www.archimuse.com/mw2006/abstracts/prg_295000735.html (accessed November 5, 2007).

38. George A. Miller, "The Magical Number Seven, Plus or Minus Two: Some Limits on Our Capacity for Processing Information." *Psychological Review* 63 (1956): 81–97, at www.well.com/user/smalin/miller.html (accessed November 5, 2007).

39. S. H. Ham, "Cognitive Psychology and Interpretation: Synthesis and Application," in *The Educational Role of the Museum*, ed. Eilean Hooper-Greenhill, 2nd ed. (New York: Routledge, 1999), 161–71.

Designing for Mobile Visitor Engagement

SHERRY HSI

Handhelds have been used to support museum learning through object interpretation, exhibit guidance, and wayfinding, much like traditional audio tours. With new multimedia features, more computational power, and communication capabilities, handhelds can offer visitors a broader range of access to informational, social, and cognitive supports, depending on their design and application. Larger devices, including wireless tablet PCs, portable CD players, and notebook computers have also been used as media communication devices to demonstrate phenomena, to notify visitors of ongoing events, to foster dialogue and messaging between the visitor and the exhibit designer, or to support discussion between docents and the floor staff. Used as documenting and sensing tools, handhelds combined with probes and radio frequency identification (RFID) can collect data, monitor events, and capture digital photos to extend a museum visit with analytic and constructive sensemaking activities. Other applications using visitors' own cell phones and personal digital assistants (PDAs) can engage them in playful educational group games that simulate or model particular phenomena, thus personalizing the experience.

These technology designs all have a common thread—using the museum space as an innovation testing ground for orchestrating and configuring different media and technologies to provoke new thought and interaction. However, the introduction of technologies into museums, like most high-risk experiments, can also sometimes become expensive ventures with unintended outcomes. The fundamental and recurring issue is how to design a museum environment with a palette of people, exhibits, media, technology, and objects into a compelling and memorable educational experience. Given the multiple pathways possible, each visitor's museum learning experience is highly personal, so it can be difficult to support all visitors with technology tools that tend to be designed as one-size-fits-all applications.

The goal of this chapter is to provide insights and reflections upon a decade-long history of handheld and networked technology design experimentation, collaborative partnerships, and evaluations that have been conducted at the Exploratorium with visitors, teachers, staff members, and docents ("explainers") from 1997 to 2007. How have different technologies, including handheld computers, high-speed networking, RFID, wireless computing, and interactive exhibits, been designed to support on-site user experiences and off-site extended learning? From these designs, what are users telling us about their preferences and experiences? In reflecting upon this work, this chapter highlights the design challenges and trade-offs that museums face when considering full-scale implementations of wireless and mobile handheld technologies. Their implementation in museums poses a range of complex and interesting cultural, social, and pragmatic design issues at the individual, programmatic, and institutional level. Throughout this chapter, hard-won design insights are offered to help inform future museum technology applications during the conceptual design phases, to ensure user experiences and context-specific affordances are included in social design process.

INFORMATION LANDSCAPES

A first-time visit to the Exploratorium can easily be exciting but also overwhelming, due to the colorful sounds of conversation and discussion intermingled with the noise of fans, bells, bubbling water, spinning disks, marimbas, motors, and pipes. The museum houses over four hundred unique interactive exhibits and working prototypes that engage the visitor with a particular scientific concept or perceptual phenomena. Some exhibits have the appearance of ongoing science experiments, while others take the form of curious, awe-inspiring art pieces. To make exhibits more accessible, "explainers" (who work at the museum as docents) serve as intermediaries between exhibits and the public to help facilitate a deeper understanding of the science behind the exhibit, promote exhibit play, and in general help visitors. In this rich and complex social learning environment, any introduction and application of technology would need to be carefully considered and crafted with the same intentionality as other exhibits and programs so the design could fit well into the personal and social nature of discovery at the museum and to support curiosity, engagement, and learning.

To help visitors and staff members make connections to and between exhibits, the Exploratorium created a learning resource called the exhibit cross-reference. This initially began as a staff-only resource in 1975 to catalog exhibits using punch cards to record different information relating exhibits to scientific and perceptual phenomena that could be later printed out on paper. The cards and paper were simple tools for instructional mediation. However, the paper lists were stored in file cabinets and eventually lost. The cards were cumbersome to use at exhibits for teaching and difficult to maintain. This metaphor for linking ideas to bring coherence to information and artifacts across a physical landscape persisted as a design concept, even as technology evolved from punch cards to hypertext and the Internet flourished (and continues to grow) from desktop to ubiquitous wireless media access.

As an example of a reification of this concept, a technology-assisted experiment was conducted in 1998 called "Ghost Landscape." The innovation lay in its use of small, ubiquitous computers and wireless networks to allow the integration of an outdoor setting with a mediated, abstract world of information, making possible a new informal inquiry-based curriculum for teaching ecology. Visitors walked through a real landscape outside the museum with a networked tablet PC, linked to a carefully selected set of information and media related to their direct experience of the ecosystem.

The curriculum operated in two parallel realms in order to create a bridge between abstract information and concrete experience. The first part, the concrete, is an interpretive nature trail in the restored Crissy Field wetlands, located in San Francisco's Golden Gate National Recreation Area, near the Exploratorium. The information overlay, the abstract, is a Web-based database of text, images, video, and curricula about the natural ecosystems in and around the wetlands. The online information is navigated through a visual representation of the trails and of the wetlands at large; at the same time, sensors in the environment read the movements of the visitor, enabling the delivery of information specific to the location. Thus, the virtual landscape becomes a map of the real landscape, an overlay of information and media to be explored by the visitor, to whatever depth or level of complexity the visitor chooses. While either site may be used independent of the other, each is designed to illuminate and locate the other through physical pointers and digital representations.

THE STORIES: INFORMATION AS A GHOST OF THE LANDSCAPE

Focusing on the lagoon surrounding the Palace of Fine Arts (the first home of the Exploratorium), the museum staff collaborated with the neighboring national park organization and a local group of radio enthusiasts led by Brewster Kahle (founder of the Internet Archive). These partners provided, respectively, a source of educational content and a 2MB per second minimum throughput in the existing Presidio wireless network. This network, considered fast at the time, allowed full multimedia to be delivered to the user as part of the information mapped on to the Crissy Field wetlands landscape. Staff members gathered, recorded, and created a corpus of information about the geographical space. This information was multidisciplinary, including the following:

- Ecology: Information on the plants, animals, insects, fish, and birds that inhabit the area and their complex interrelation to each other. Most of this information was acquired and shaped through work with City Park staff.
- History: Information on the construction and life of the lagoon and its surrounding architecture and its relation to the city of San Francisco and the Exploratorium. This information was the product of research into Exploratorium and San Francisco city archives.
- Art: Many visiting artists at the Exploratorium have used the lagoon as a site for art installations, sculptures, and performances. Written, still photo, and video records of these works have been kept by a variety of Exploratorium staff members, and this project enabled the collection and organization of these records.
- Memory: Personal and institutional memories, stories from the life of the lagoon, the Exploratorium, and its staff. Much of the rich history of the lagoon for Exploratorium employees involves the stories and events that have happened there, ranging from staff members wading into the lagoon to maintain an art piece or rescue a snapping turtle, to the time a car careened into the lagoon.

An outdoor evaluation found that the areas where the science content overlapped with personal stories and anecdotes seemed to have the most interest for visitors. From testing with visitors and staff members, our first design insight emerged about creating content on handhelds for learning.

Insight 1: Content should be socially relevant, educational, and actively personal. Users expressed interest in being able to add personal stories to the Web resource through the form interface, so that the information that hovered "ghostlike" about the lagoon would reflect not only the scientific and academic history of the place but also a growing human history, the "ghosts" of those who had been there before. Thus, the interface allowed a link to an e-mail form for users to submit their sighting of fauna, or their question or thoughts about the other topics presented. This was a useful approach to collect distributed observations and memories to share, but no dialogue was established because, as an experimental project rather than an ongoing program, no person was dedicated to answering these e-mails. (This echoes similar institutional constraints mentioned in the previous chapter.)

Given the form factor of the technology at the time, weight and outdoor visibility were the biggest complaints from users. The tablet computer weighed over four pounds, and despite its better multimedia and display technology, users reported fatigue in carrying it around. The evaluation pointed to the need for devices to be lightweight with powerful color displays if users are to have viable access to this kind of information outdoors. The networking technology, on the other hand, worked flawlessly, offering robust, fast connections to the network that completely met the demands of the context of use.

Insight 2: Select technology that matches the physical context of use so it becomes seamless with the user experience.

Electronic guidebooks: Encouraged by the positive reception of the content-based mediation, our next set of experiments involved refining and extending further the idea of an information landscape by fostering more personal inquiry and exploration, as well as experimenting with different media types afforded by lighter weight and computationally more powerful technology. The driving metaphor that shaped the kinds of interactions desired was a paper travel guidebook. With this application, visitors could look up interesting nuggets of information about an exhibit, scribble personal notes, take souvenir photos, learn about the history of who created exhibits and how

FIGURE 8.1
The Hewlett-Packard Jornada weighing in at 1.1lbs (520g)

exhibits were designed and changed over time, and peruse suggestions for interacting and teaching with exhibits.

The Guidebook experiment was marked by a fruitful collaboration with the Concord Consortium, who brought inquiry design ideas, handheld research expertise, and probeware experience, and HP Labs Palo Alto,[1] which brought technical expertise, researchers in human-computer interaction, and hardware in the form of RFID and infrared communication technologies called "beacons." The museum provided expertise in content and media development, exhibit design, informal learning, and visitor evaluation. Several portable computers were considered, and a clamshell-sized computer was selected because it had best ratio of screen size to weight at the time. The latest wireless protocol enabled digital video and audio to be delivered, as well as messaging to a server to create two-way communication and information-sharing experiences.

Rather than retrofit existing exhibits, stands called PI-stations were built with a computer laptop base to mount different identification technologies including wired RFID, infrared beacons, and a networked eyeball digital camera to enable exhibits to be responsive to visitor interactions with handhelds.

The station allowed the exhibit to recognize the presence of a visitor, uniquely identify them, and display exhibit-related content to appear automatically on the user's screen when she or he came within proximity of an exhibit.

The multimedia content was designed by staff members to include video-only, audio-video, and audio-only tracks. The first content page shown on the handheld for a given exhibit always encouraged the visitor to engage with the exhibit so they would not get overly engrossed with their heads down in the device, detracting from the social experience of museums (see figure 8.2). The

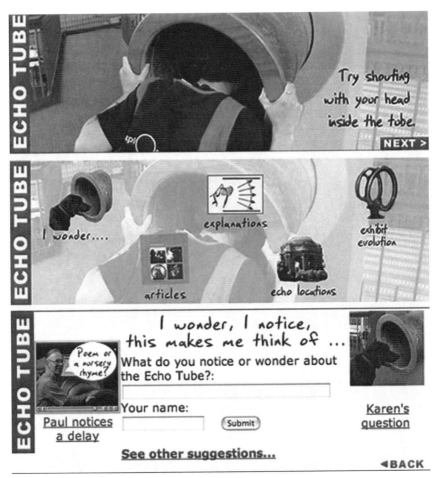

FIGURE 8.2
Three screens from Electronic Guidebook implementation on HP Jornada 720 clamshell computers.

second-level page organized content into "nuggets" that included historical articles from the museum's print magazines, historical stories, explanations of science, and everyday places around the San Francisco Bay Area where other "echoes" (related information) could be found. The handheld supported not only interpretation and explanation but also the idea of "remembering" a visit to the museum by allowing visitors to bookmark the name of the exhibit they were visiting and sending user-captured souvenir photos from the PI-station to a Web page for viewing later back at home or school.[2]

At the start of the project in 2002, there was no institution-wide wireless network plan or implementation at the museum. The wireless capabilities that existed were enabled by individual museum departments and projects that purchased wireless hubs to provide local coverage that was sufficient for purposes of testing, but this limited the locations where users could access the guidebook content. While mobile phones were considered, even though their media delivery capabilities were limited at the time, the coverage was spotty in the Palace of Fine Arts (home of the Exploratorium), and a partnership with a commercial service meant limiting users to use audio or text for mediation in a proprietary format.

In the evaluation of this guidebook, the results showed a mixed outcome— while it successfully introduced visitors to new information about exhibits and prompted play with exhibits in new ways, visitors did not like to carry handheld devices while visiting the museum. As with the Ghost Landscape experience, they found the device to be distracting and awkward to carry, even if given a pouch in which to carry it. Unlike the outdoor Ghost Landscape, this indoor experiment meant having the user attention split between surrounding people, the hands-on experience of the physical space of exhibits, and the online experience, even if the content necessarily prompted the user to play with exhibits.[3] While it was thought that helping the visitor find content with automatic cuing would help, this often surprised the visitor or interrupted the flow of reading.

Like the Ghost Landscape, visitors could contribute questions or topics via the interface, but the competition for attention in the physical space of a museum, combined with the awkwardness of one-handed typing, discouraged use. In addition, visitors told us they did not want to enter in annotations into what seemed like a personal device because it was not clear where those comments would be posted. This outcome might have been different if the users personally owned the handheld communication devices.

Insight 3: Online messaging succeeds when the design makes explicit when messages are private or public, where they are posted, and who reads them. On the other hand, the photos taken by visitors themselves, triggered at PI-stations using their handhelds, while limited in quality because of poor lighting and camera focus range, were found to be engaging to visitors, helping them remember their visit. The inherent nature of wanting to view personal media encouraged more than half of the visitors who tested the device to visit a Web page after their museum visit—a rare departure from the low levels of website follow-up mentioned in the previous chapter.[4]

Guidebooks were successful in offering the visitor new suggestions and background information on how to manipulate the exhibits, especially for the more timid visitor. While the guidebook did successfully impart new information about exhibits and prompt visitors to play with exhibits in new ways, headphones were required to listen to the multimedia content, causing some discomfort—especially for seniors with hearing aids. For some, the handheld became the primary exhibit as the visitor became engrossed in watching multimedia videos, reading explanations, and tapping on the screen through pages. For most, the size and weight of the device with headphones detracted from the social dialogue and learning conversations that are some of the hallmarks of learning in the Exploratorium.

Based on the Guidebook interviews, it was clear that many users—especially teachers—wanted more information about exhibits, but only after their visit to the museum. Explainers, on the other hand, had an easier time multitasking and attention switching: watching a digital video guidebook demonstration while conversing with other explainers, talking to visitors, and playing with exhibits. Explainers were inherently interested in the content because their job in the immediate context of use required them to learn how to teach with exhibits. Two more insights emerged from our work with guidebooks:

Insight 4: Multimedia content can be designed to be compelling enough to provoke users to try new things with exhibits, even in a setting that competes for social and physical attention.

Insight 5: Different audiences need to be studied early on to help situate the handheld into their work and flow of activity.

The staff went back to the drawing board, this time focusing on the three promising areas of development: extending museum experiences for visitors without handhelds, personalization, and explainer handheld applications.[5]

EXTENDED LEARNING THROUGH
PERSONALIZATION AND EXPERIENCE CAPTURE

The physical environment of the Exploratorium provides almost too many things for the user to do. Given that we wanted to foster learning and we had prior success with visitors using the "remembering" photo capture feature of the first electronic guidebooks, we explored what kinds of interactions would allow the on-site learning time in museums to be extended beyond the museum visit—into the periods of time both before and after a visit to the museum—as John Falk and Lynn Dierking suggest in chapter 2.

A team of researcher, graphic designer, exhibit developer, and science writer generated paper-based scenarios of use for each audience group to identify the flow of activity and evaluated them with visitors, teachers, and explainers for verification, feedback, and refinement. Each scenario presented activities to perform before, during, and after museum visits.

In the museum visit scenario, we imagined the user having a "smart watch" that would recommend exhibits to visit, record which exhibits a user had

FIGURE 8.3
Paper-based scenario to brainstorm applications and flow of user activity

enjoyed in previous visits, serve as remote control to trigger a digital camera, or act as other types of sensing probes for personal data collection.

In addition to the paper scenarios, which stressed the sequence of events during a visit, we did a number of paper prototypes for the screen interface that visitors would use after their visit to the museum. For example, in one set of activities, a visitor could use a Web-controlled video microscope probe to look closely at magnetic bacteria in the lagoon by the museum, with text or video comments from the point of view of a biologist or a physicist. A photograph taken from the museum's pendulum drawing board exhibit or heat camera exhibit (exhibits that allowed users to create something on-site) could be automatically posted into a Web page, and the online activity would allow further exploration in an online drawing of patterns and pendulum simulation activity, or heat inquiry activity. These were tested in focus groups to better understand what kinds of simple tools and activities a user might want to engage in during their "post-visit" to a museum. Based upon common themes that arose from focus groups, we identified the following interactions: build, interact, share, collect, and remember a visit to a museum. These led to a set of common tools that could be used for any given activity in a digital notepad. The notepad had a personalizable launch pad that contained the tool kit and links to other personal media.

eXspot Wireless RFID Reader and Bookmarking System

It was time to translate the paper-based designs into a visit-extension prototype. Web technology can extend the museum experience by allowing visitors to access content both before and after visiting, but findings based on exit surveys showed that less than 1 percent of visitors went to the museum's website before coming to the Exploratorium, even if they were regular Internet users. Teachers were more likely to prepare field trip-going students for a visit beforehand. Few visitors knew that the Exploratorium even had a website. Regional visitors knew we had a site but used it only to find out hours and directions to the museum.

When asked about extending their experience, most visitors liked the idea of in-depth content and activities being available to them but said they would prefer to access such resources after a visit on the Internet. Therefore, the moment offering the highest potential for deeper learning from a visitor's perspective was deemed to be after the visit. Based on our prior experience with

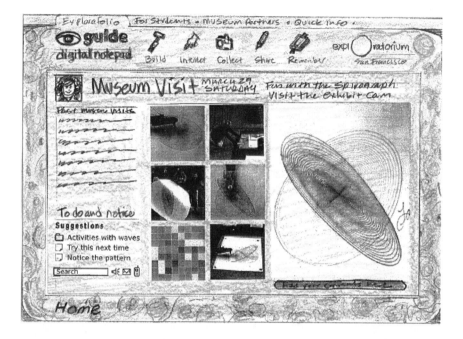

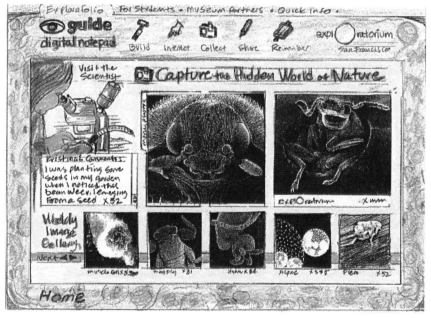

FIGURE 8.4

Paper prototypes of post-museum activities that linked the museum visit and the media created at the museum with an online activity, such as the Drawing Board or the Microscope Imaging station.

the electronic guidebooks, we knew we needed a method for recording and capturing the user experience that was "lightweight" enough (both literally and cognitively) such that it only minimally intruded on the physical experience at the museum. Because of this, we investigated a variety of alternative ways of recording and capturing the user experience, such as remote clickers, smart watches, and smart cards.

The eXspot project revisited the use of RFID technologies to address this problem by allowing users to "tag" exhibits that pique their interest during their visits and learn more about them online after their visits. At that time, RFID technology was becoming prevalent and inexpensive enough that the Guidebook experiment could be made wireless, thus eliminating the need for separate setup, like the PI-stations. Through a partnership with Intel Labs Seattle and the University of Washington, expertise, technology support, engineering design, and hardware design were contributed.

The eXspot application consists of low-power RFID "eXspot" transceivers placed throughout the museum and wirelessly connected to a logging server. Visitors receive an eXspot ID card upon their arrival at the museum and can place their ID tags on any of the eXspots to "tag" that exhibit. Their ID number is saved on the eXspot server, along with other exhibit-specific information generated by the visitor—this could be a simple photograph, a sensor reading taken by the visitor at the exhibit, or any other digitizable output of the exhibit. Like the Guidebook, some exhibits are wired with cameras to take pictures of visitors as keepsakes. When visitors return home, they log into the eXspot Web pages to view the pictures of themselves or their families at those exhibits, and they usually stick around to learn more about the science behind the exhibits they visited.

The eXspot was developed for a public audience with exhibits from the Energy and Matter Worlds. It took into account lessons learned from earlier prototypes and replaced the handheld device with RFID cards that interacted with custom-built wireless transceivers built into selected exhibits—in particular the Heat Camera and Watch Water Freeze exhibits. A computer program was written to simultaneously control and capture a video image directly from video output of the heat camera as well as from a camera pointed at the visitor. These were both sent over the network to a private Web page created expressly for that visitor after an eXspot card was used to trigger the camera. Back home, visitors could then scroll between their real photo and the heat

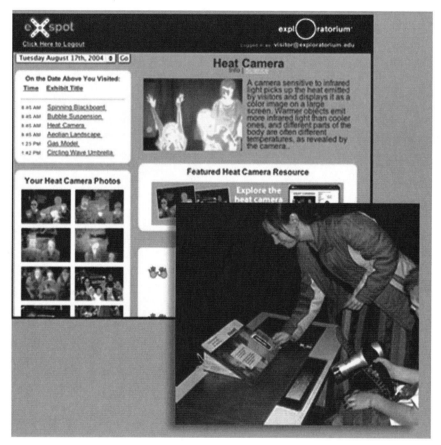

FIGURE 8.5
A user at the heat camera exhibit using eXspot to capture an image, and a screen of the post-visit website with heat camera images.

camera image of their photo. A wireless tablet was used as the heat camera exhibit label, to provide a countdown to the digital photo capture as well as to briefly display the results of the digital photo capture. Similarly, an experiment was done with the Watch Water Freeze exhibit, except that in this case photos captured by the visitor using the eXspot RFID system depicted water crystals freezing.

Related to our first insight about experiences needing to be actively personal, another insight regarding the technology application emerged from our eXspot experiment:

Insight 6: Designing personalization in technology can be a powerful lure. People are inherently interested in themselves and discovering something interesting about themselves. Personal media created by a visitor can act as a lure, to both try a physical exhibit and check out media after heading home.

While this system eliminated the need to carry around a handheld device, it introduced a relatively new and unfamiliar technology. Even though most visitors interviewed were relatively computer savvy, pilot studies showed a need to educate people about RFID technology. Some visitors were concerned with privacy issues, but most lacked a basic understanding of how information was transmitted to and from an RFID card and how information could be made secure over a wireless network. We concluded that visitors needed to learn more about how RFID works before being able to fully adopt and use it successfully. Plastic cards that were used to trigger the RFID reader to bookmark a visit were developed to explicitly show the antenna and chip on the card and give a short explanation about RFID, as well as some signage and handouts during the testing. This helped provide some education to the public and help belay some of the fears around RFID. Ideally, an orientation station with a few exhibits demonstrating how RFID works would be useful in future permanent installations, at least until RFID is a commonly used technology.

Insight 7: Using cutting-edge technology creates overhead for the experience. Whenever brand-new technologies are employed, a certain amount of visitor education is needed before they can use the new technology. This should be seen not as a mark against experimenting with new technology but rather as another education opportunity for museums to perform their mission to educate the public—having used a new form of technology, a visitor returns to the outside world with a broadened understanding of what is possible with handhelds and technologies like RFID. This is a trade-off using novel technologies, but if used, an orientation and educational ramp-up should be planned for, both in terms of time and in terms of an exhibit or ancillary materials.

NETWORKED DISCUSSIONS AND DIALOGUES

After seeing their enthusiastic use of the previous guidebook, we set forth to develop a specialized application for museum docents ("explainers," who are high school-aged and young-adult docents) that would more directly support their role in the museum.

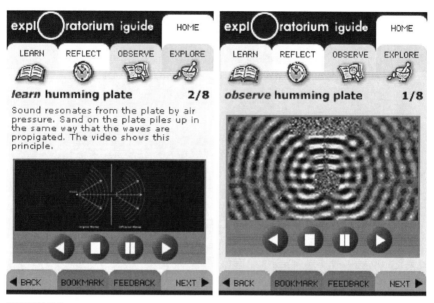

FIGURE 8.6
An example of an electronic guidebook user interface prototype designed for explainers which was abandoned in favor of a peer-to-peer communication application, the Q&A.

Using the "Q&A" prototype's mixture of multimedia and interactive discussion group-bulletin board functions, the museum's explainers could help their peers answer each other's questions while learning on the job how to be an explainer. The design of the Q&A application followed after doing multiple focus groups with explainers and showing them various prototypes of handheld applications. Much of the early prototypes were heavily content based, but the docents expressed more interest in the communication capabilities of the devices: being able to communicate with each other about real-time problems and information was more valuable to them than content alone. This application worked well to combine multiple pathways to access the discussion: on-site handhelds, home computers, museum computers, and Internet-enabled phones could all access the conversations.

There exists a wealth of knowledge among explainers and their teachers concerning scientific phenomena about exhibits, as well as tips about how to teach these phenomena in an informal science learning environment. However, not all explainers are able to maintain or collect this knowledge, due to the decentralized manner in which it is dispersed. Q&A was designed to address this problem and improve the understanding explainers had about phenomena, as well as to improve the communication between explainers on the museum floor. Explainers with handheld computers on the museum floor could distribute questions and answers about exhibits and phenomena, as well as access a log of past questions and answers for reference. Explainers use the handheld to enter text questions, then post them using the Exploratorium's wireless network. The question is then posted on a Q&A page for other explainers to view using their handhelds (or home computers, phones, etc.). Explainers can post answers or peruse previous questions and answers in order to expand their knowledge about an exhibit or phenomena, or to get pointers on their presentation and interaction with guests.

Insight 8: While handhelds are good at knowledge delivery, user-generated information on handhelds is more socially relevant and interesting. Explainers, who are also younger and more tech savvy than the typical museum visitor, were able to generate information for a handheld that was immediately useful in their community. The information posted could be attributable to someone in the community. Thus, the handheld was viewed not institutionally as a threat of being a docent replacement but rather as an educational support tool. The success of the Q&A application also rested on enabling real-time, nearly synchronous communications among museum staff members. We continued to search for a role for technology that would enhance the real-time social experiences of visitors to museums, rather than supplant them.

NETWORKED ACTIVITIES AND GAMES

One promising usage paradigm for handheld devices, called "participatory simulations," caught our interest because it is designed to encourage predominantly social, face-to-face, real-time interactions. Previously explored in classroom contexts, participatory simulations use handheld devices to mediate a role-playing game that commonly centers on emergent phenomena like disease transmission, economic trading, or trait inheritance. Each player assumes a role (e.g., a fish with a specific genotype), and the handhelds help mediate transactions (such as mating) with other players that result in changes to the overall state of the population (e.g., the distribution of phenotypes among the different players). Unlike several of our earlier experiments, social interactions are foregrounded with the technology playing only a supporting role.

Today, more people—especially teens and young adults—carry handheld technology with them and use them fluently as part of their everyday life. Given the computational and communication capabilities of these devices, we revisited the idea of allowing visitors to use their own personal devices to join real-time collaborative activities designed and provided by museums, now that mobile service providers are more open to cross-platform development. Such collaborative simulations match well with the social nature of learning in museums, but there remain many open design questions.

To address the issue of social isolation with multimedia PDAs, we experimented with a handheld application that was originally designed for school users, called MUSHI: Multi-User Simulation with Handheld Integration.[6] This framework is similar to traditional participatory simulations but has the added element of a large, shared display that illustrates the current state of the shared simulation. The software that was tested is a Game of Life-style simulation of cancer growth in human tissue. Each participant assumes a different role (e.g., surgeon, oncologist, or radiologist), which is supported by the individual devices: each player makes use of an interface that is tuned for the actions allowed to their given role. The large, shared display graphically depicts the status of the patient, so the players can see how each of their individual, specialized actions impact the patient's health. They are tasked with working together to try to eliminate the cancer without killing the healthy tissue. Each user can provide input to the shared simulation of cancer growth via their individual handheld devices.

Since the purpose of introducing this kind of activity to museums was to provide a social learning experience, our experiment with this prototype focused on studying the impact that different ways of providing information to participants had on the degree of participants' collaborative activities. While dialogue is an important indicator and data source for collaboration, we also carried out an observational study to look for the presence or absence of behaviors that indicate cooperation (like productive dialogue, task sharing, planning, and participation by all group members). A potential pitfall of participatory simulations is that the learner spends more time collecting and interacting, at the expense of time spent on reflection and understanding the large phenomena that emerges from the simulation. At the writing of this chapter, this research is ongoing.

REFLECTIONS AND DESIGN PRINCIPLES

Over the past ten years, handhelds have served in different roles and capacities as information landscapes, guidebooks, discussion tools, participatory simulation games, and other extended learning applications in and beyond museum walls.

In all these design experiments, design began with a focus on an existing or desired informal learning and user experience, and was shaped by a strong guiding metaphor for the experience, although these metaphors differed between applications. Sometimes the guiding metaphor was to support dialogue directly with another user, as in the case of the explainers' Q&A application. In other cases it was to support indirect dialogue; in these situations, the application was to serve as a secondary index to another user via their voices, knowledge, media, stories, questions, photographs, and historical voices (aka "Ghosts").

Successful transformations came from participation and collaboration by different design partners, evaluators, and community stakeholders, all of whom worked to contribute resources, expertise, and critical dialogue on issues in order to further inform the design of content, interactions, and activity intended to take advantage of physical setting, the mobility offered by technology in museum spaces, and resources and people in that setting. Museums could not have done this work by themselves.

Insight 9: Actively search for synergistic collaboration and partnership opportunities to support capacities not already in museums. Any museum technology project intending to have a lasting impact will be plagued by pragmatic issues including the management and sustainability of the content and technology. Technology becomes quickly obsolete, and maintaining a cyberinfrastructure in a museum is often not a priority, compared to operation of day-to-day programs. Museums need to examine ways to support infrastructure and seed the formation of technical assistance. One can imagine that, with ubiquitous wireless and cell phone coverage, industry in partnership with communities and government will be responsible for maintaining reliable technological infrastructure, while museums can focus on the visitor experience—human interaction, content for interpretation, and media design.

Insight 10: Create designs that encourage and promote broad participation from different audiences. Museums as public-serving institutions face the issue of inclusivity—ensuring that a broad participation from different and diverse visitor audiences can benefit from its offerings, as Ellen Giusti explains in chapter 6.

In our experiments, we considered designing for different groups of users but also wondered if adding technology to their visit would be an additional burden, detracting from direct access to exhibits and educational programs. Groups such as seniors, urban youth, migrant families, and multigeneration and multilingual families pose new design opportunities, especially as multimedia content can become more dynamically generated to address the diversity of audiences, and technology can be better designed for a seamless user experience rather than an add-on. This is an exciting area that has yet to be explored.

Handhelds, like any other new technology innovation, will have a higher rate of success if they fit into the culture and philosophy of the local institution. For example, a PDA that is didactic will not survive in an environment designed for learning through social conversation and exhibit play. Institutions that have a strong immune system response will attack and weaken the success of adoption. Key questions should be asked before the start of any technology design for museums, some of which are articulated in figure 8.7.

Museum spaces offer a range of environments from noisy and hands-on to quiet and contemplative. Learning with the use of handhelds should fit into the context of use, whether a social environment or a private, individual experience. Handhelds can permit the design of a personally relevant experience that can be customized to a person's interests. The handheld is not just a mini personal computer but should be orchestrated and situated into a whole set of human interactions and experiences that are actively personal.

- Does the application or designed experience fit the culture and philosophy of the place?
- Does the application make the user experience actively personal?
- Is it intended to be a one-time provocative experience or an extended learning opportunity?
- What is the life and duration of the experience? Is it intended to be an ephemeral experience or a persistent extended learning experienced shaped by many individuals over time?
- Does the application need to fit into an existing program?
- What does the user experience, and how are their needs being addressed?
- What is the commitment of the organization or institution in supporting this technology, including the technology infrastructure?

NOTES

1. HP Labs Mobile Ubiquitous Research Group and CoolTown Beacons.

2. Margaret Fleck, Marco Frid, Tim Kindberg, Rakhi Rajani, Eamonn O'Brien-Strain, and Mirjana Spasojevic, "From Informing to Remembering: Deploying a Ubiquitous System in an Interactive Science Museum," *Pervasive Computing* 1, no. 2 (2002): 13–21.

3. Sherry Hsi, "I-Guides in Progress: Two Prototype Applications for Museum Educators and Visitors using Wireless Technologies to Support Informal Science Learning," in *Proceedings of the 2nd IEEE International Workshop on Wireless and Mobile Technologies in Education*, JungLi, Taiwan, 2004, 187–92.

4. Mirjana Spasojevic and Tim Kindberg, "Evaluating the CoolTown User Experience" (paper presented at the Workshop on Evaluation Methodologies for Ubiquitous Computing held at Ubicomp'01, Atlanta, 2001), at www.exploratorium.edu/guidebook/eguide_exec_summary_02.pdf (accessed November 6, 2007).

5. This work was supported by a grant called iGuides from the U.S. National Science Foundation.

6. Leilah Lyons, Joseph Lee, Christopher Quintana, and Elliot Soloway, "MUSHI: A Multi-device Framework for Collaborative Inquiry Learning" (paper presented at Proceedings of the 7th International Conference on Learning Sciences, Bloomington, Ind., 2006), 453–59.

9

Cross-Context Learning

PAUL RUDMAN, MIKE SHARPLES, PETER LONSDALE,
GIASEMI VAVOULA, AND JULIA MEEK

Learning from a museum visit need not be confined to the brief time actually in the museum. With the development of new personal technologies there is an opportunity to offer an extended learning experience, one that connects the context of a museum visit to activities that take place before and after it, and that allows a visitor to communicate with experts, tutors, peers, and him- or herself over time.

To design a learning experience that extends throughout and beyond the museum, we need to place the learner at the center of activity, interacting with whichever technology is appropriate for the situation. For example, a learner may capture an image of a museum exhibit using a camera phone, then return home to study the image further, along with text from the museum's website, using a desktop PC. The same image may be viewed again on a classroom computer as part of a school project.

Putting learner mobility at the center of the analysis illuminates museum learning from a new angle. It allows us to explore how knowledge and skills can be transferred across contexts, such as between school and home, and how new technologies can be designed to support a society in which people on the move try to cram learning into the gaps of daily life.

Thus, a key issue is to understand how people artfully create impromptu contexts for learning. A parent may use an exhibit in a science museum to create a "teachable moment" in order to explain why tides occur, or two people may meet in front of a picture in a gallery and discuss the painter's use of texture. The same painting may raise questions in the minds of visitors, causing them to take time out from the bustle of life for reflective learning—to look around and ponder.

Rather than treating the context of learning as a fixed setting that a person enters and leaves, we can reconceive it as an intrinsic part of the learning activity that is created as the learning progresses. The common ground of learning is continually shifting as we move from one location to another, gain new resources, or enter new conversations.[1] A challenge, then, for designers of mobile learning technology is to support the creation of temporarily stable sites for learning, providing appropriate technology and services at just the time and place needed to support productive interactions and access to learning resources.

One solution is to provide people with general-purpose portable devices that can easily be carried and used in any circumstances, regardless of location. However, the learning should not rely solely on handheld devices but can be formed from the particular setting, the available artifacts, and people whose knowledge and interests are continually changing. How can new technology support useful learning in context?

One solution to this problem is to design technology that provides context-dependent content and services. On a basic level, a system might deliver content to museum visitors based on profiles of the learners, their current locations, whether they had previously been at the locations, and how long they had been there. On a more sophisticated level, it could furthermore take account of their trails of movement, the history of their learning, and the available resources, including experts and other learners. Although people could share the resources—for example, a couple walking round a museum—this is in essence an approach to providing personalized learning through context.

An alternative, though not necessarily conflicting, approach is to support mobile learning as a collaborative activity. Here, the focus is on not the learners, nor their technology, but rather the communicative interaction between them, in order to advance knowledge building. We claim that conversation drives learning.[2] It is easy to overlook the power of being in conversation with others.[3] Conversation is the means by which we negotiate differences, understand each other's experiences, and create consensual interpretations of the world. This is particularly important in a mobile context: a 2005 study of everyday adult learning, based on personal learning diaries, found that the biggest difference between mobile learning and fixed learning was in the use of conversation. In 45 percent of mobile learning episodes, conversation was used as a resource for learning, compared with 21 percent for nonmobile.[4]

The obvious role for technology in collaborative learning is to provide a channel for communication. A student phones a helpdesk to ask for advice on how to fix a computer; a doctor contacts a colleague to discuss a diagnosis; a researcher e-mails a colleague to ask for a journal reference. Kevin Walker gives other examples in chapter 7.

However, this view of collaboration as an exchange of knowledge, over an inert and transparent conduit, does not embrace the growing use of technology as an active medium for creating and sharing knowledge. For example, Wikipedia does not merely convey information, as one would find from a conventional encyclopedia website; it is also a dynamic medium through which people coordinate their understandings by engaging in a continuing dialogue.[5] This can either be implicit, through the shared enterprise of constructing and modifying the online encyclopedia, or explicit, through an online conversation associated with each Wikipedia entry.

This form of collaboration, along with technology-mediated social networks, is now mobile. For example, Google Maps is available on mobile phones,[6] and photo-sharing sites such as Flickr link to maps so that people can take photos on the move and make them available along with location details.[7] People use mobile devices to become active participants in a continuing conversation: leaving voice messages, sharing contact lists, creating online diaries (blogs), and capturing and sharing images.

Learning, then, is a continual conversation with the external world and its artifacts, with oneself, and with other learners and teachers. The most successful learning comes when the learner is in control of the activity, is able to test ideas by performing experiments, and is able to ask questions, collaborate with other people, seek out new knowledge, and plan new actions.[8] In addition to these external conversations, each learner holds a continual internal dialogue, making sense of concrete activity by mental abstraction and by forming theories and testing them through actions in the world.[9] Learning is driven by the interplay between internal and external conversation, mediated by the changing context and by external shared media.

There are many roles for technology in such a vision of learning. It can make connections, by such means as phone calls, voice and text messaging, blogs, discussion websites, and location-based information. The connections may be created between physical locations, between people with differing views and expertise, or across time. Technology can also act as a conversational partner, offering information related to learning within the immediate context and providing the means to exteriorize agreements through shared notes, maps, and diagrams.

CASE STUDY 1: GEORGE SQUARE

Background

As discussed, learning is a continual conversation with the external world and its artifacts, with oneself, and also with other learners and teachers. The concept of the "external world" includes the obvious physical spaces and objects, from buildings to possessions to artifacts in a museum. It also includes representations of these physical objects, external to the learner's mind. Since these representations need not take the form of the object they represent, they are easier to create and manipulate—for example, a drawing on paper of a building, or a written name instead of a drawing. They may also represent nontangible ideas and concepts.

Traditionally, such representations are often used in conjunction with real-world objects. A map is used while walking around a city space; a worksheet is used by students in a museum. In today's technological world, representations may take many forms—not just text, diagrams, and images but also sounds, animations, and video. It may even be that a situation is better understood through its representation. Can one best experience a new city by walking around the space or by playing a computer game set in that city? What is the best way to discuss and compare museum exhibits, by walking between them in the museum or by manipulating photos and descriptions of them on a personal computer?

The Equator project[10] in the UK explored the contemporary relationships between the physical world and its digital representation, and how their combination may lead to new forms of experience. Within Equator, the University of Glasgow led the "City" project, an investigation into the relationships between the physical city and its digital representation.

Description

The George Square system helped people explore the main public square, George Square, in Glasgow.[11] Pairs of visitors, who already knew each other, were asked to explore and learn about the space: its artifacts—statues, monuments, memorials, and so forth—and history. One visitor was free to walk around the square, while the other sat in a nearby café. Both had computers, tablet, and laptop respectively, giving access to digital representations of the square in the form of a map, photos, and Web pages. Both could take and share photos; both could see each other's location on a real-time map (GPS location from the square, manually selected location from the café). Headsets allowed continuous audio communication between the visitors.

The on-screen map (figure 9.1) showed the visitors' locations, photos taken, and Web pages viewed in the related location. In addition, previous visitors were represented. The paths the previous visitors traveled and the sequence of places visited were stored on a database and matched against the current visitor's path. The matching stored path was then followed forward to create recommendations based on other places visited by the earlier visitors, photos they had taken, and Web pages they had viewed.

Thus, both visitors could talk to each other, browse the Internet, take and share photos, and see each other's locations and recommendations on a real-time map. Only one person, however, was actually in George Square.

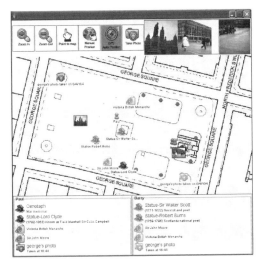

FIGURE 9.1

Screenshot, showing map of George Square with each user's location, camera viewfinder (top-right), recent photos (top and as thumbnails on the map) and recommendations (lists at bottom, per user, and as icons on the map)

Discussion

One may anticipate that being physically present in the square would pre-dispose the visitor to ignore the technology in favor of engaging with "reality," while being in a warm café could set off a separate quest of Web surfing. Indeed, the person in the square did interact with the physical space by walking around, reading inscriptions, and taking photos, while the person in the café could not, and the person sitting in the café did look at four times more Web pages. However, these facts are misleading.

The visitors chose to collaborate in their exploratory learning. The audio was used extensively to conduct a continuing conversation between the two visitors, exchanging their differing experiences of the space to each built-up composite knowledge. For example, while the off-site visitor focused on the virtual representation (Web pages), they suggested pages for the located visitor to view. The person in the city space would use the taking and sharing of photos to highlight specific areas for discussion. Between them, they would create their own shared experience of the square based upon both the physical space and its digital (or virtual) counterpart.

In practice, the pairs quickly adopted the facilities available. For example, participants would often collaborate in taking photos. The person in the café would find an interesting location and suggest to the person in the square that they take a photograph, using the map to establish a common understanding of location.

In addition to the visitors' locations and photos taken, the on-screen map displayed contextually relevant information that the system identified, in the form of recommendations. These appeared as small icons of photographs taken and locations and Web pages visited by previous visitors, changing as the visitors moved around the space (physically, tracked by GPS, or logically, by clicking on the map). This offered the visitors a richer experience by alerting them to information other visitors used when in a similar context.

The system is not confined to the actual visit. As with museums, studies have shown how pre- and post-visiting can be important to the overall tourist experience.[12] In a pre-visit, tourists make outline plans for their visit and may use available representations to familiarize themselves with the physical space before arriving—for example, by studying a map to learn the overall layout. In post-visits they may use photos and collected artifacts to reminisce about the visit, learn more about what they saw, or share the experience with others. The George Square system supports this, using the digital information to create the same context as during the actual visit—for instance, the same recommended Web pages.

This project showed how the physical and digital can be combined. Rather than seeing them as alternatives (e.g., standing in front of a statue versus exploring a website about that statue), the project showed that, if combined appropriately, one can add value to the other. The exploration of digital information within the context of the physical encounter, and vice versa, enriched the overall experience. Visitors learned about the space through a conversational process of sharing ideas and experiences with each other, and with previous and future visitors, using tools provided by the technology.

CASE STUDY 2: CONTEXT-AWARE GALLERY EXPLORATION (CAGE)

Background

When someone visits a museum, the experience has a beginning, a middle, and an end. It is not just a series of isolated stops in front of artifacts. The CAGE project sought to address this by considering visitor movement within the gallery. We designed the system to make use of location and timing information so that we could deliver appropriate content to visitors using mobile devices (personal digital assistants or PDAs) in the gallery. When we reviewed the content available for the paintings on display, we found that many of them were linked in some way (e.g., a shared history) that was not visible to the visitors. By highlighting these links to the visitors, we aimed to encourage greater movement between the paintings beyond the usual linear path that most people were seen to follow. We also wanted to explore ways of making visits a more conversational experience, by giving visitors information to talk about while looking at the paintings.

Description

The CAGE project was a research trial at the University of Birmingham in 2005, part of the EU-funded project MOBIlearn.[13] The general aims of MOBIlearn were to explore how mobile technologies could be used to deliver learning content and activities to learners in a variety of different locations and scenarios, using a range of platforms and devices. Researchers at the University of Birmingham led the development of a generic context-awareness architecture for customizing the delivery of learning content and activities, based on the learner's activities and location.[14]

The context-awareness system determined the visitor's location from an ultrasound positioning system.[15] Small boxes placed around the gallery transmitted ultrasound signals that were received on the PDA. By comparing the times that the synchronized signals from different transmitters took to reach the device, the software could calculate the position of the PDA to within a meter. The position was combined with a measurement of the time visitors spent in front of each painting, to drive the context-awareness system.

During preliminary trials of the CAGE prototype, we found that the system's reliance on context to trigger content changes could be exploited by users to deliberately deliver different content when desired—physically moving around the gallery gave people a new way to interact with the digital representation of the gallery itself. These two aspects of context-aware gallery exploration—context as automatic content selector and context as user navigation tool—drove subsequent development of the system.

CAGE was deployed in the gallery at Nottingham Castle Museum to deliver audio content via a PDA and headphones, according to (1) the nearest painting (as determined by the ultrasound positioning system); (2) length of time at a painting (assumed to correlate with interest); and (3) previous time at this position (content is only repeated on request). Figure 9.2 shows how these contextual factors interact to determine an appropriate item of content to display. The painting name is derived from the visitors' location, and the appropriate level of detail to show is inferred from how long they have been in front of that painting. However, the system recorded a history of visitors' movements, starting with the last item of content for paintings that had already been seen. In the diagram, this is indicated by the movement history having a higher salience value (indicated by the + notation) than the time at the location. In other words, when the movement history suggested a higher level of detail, this overrode the current time at the location.

As a comparison with the "regular" visitor experience, a control group was given a printed guide and asked to take part in the same evaluation as the visitors who used the PDA. We also observed the behavior of visitors to the gallery who received no information at all (the normal experience for this gallery), to give us an indication of how using either the printed guide or the PDA would influence activity in the gallery. To measure what visitors learned during their visit, people in the pamphlet and PDA groups were asked to complete a short quiz both before and after their time in the gallery.

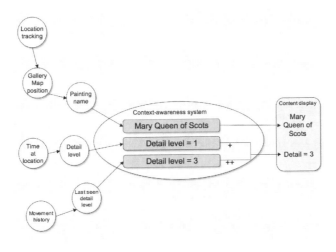

FIGURE 9.2
Flow diagram of CAGE's context-awareness system, showing interaction of contextual factors to determine content display.

Discussion

In general, people responded favorably to using the system. The use of context awareness (location plus time) to drive content delivery in this setting
worked well. Visitors were able to move around the gallery, receiving content
at the right time and at the right level of detail. The mechanism for delivering
more detailed content seemed to fit well with visitor expectations—feedback
indicated that visitors were not surprised by what they saw and that the content displayed was often judged as being appropriate and useful.

One problem observed was when visitors who quickly developed high expectations of the system (e.g., by seeing a large number of content items for
some paintings) were subsequently confused and disappointed when other
paintings had fewer content items available.

The evaluation of what visitors learned during their time in the gallery suggested no significant differences between the PDA and pamphlet as sources of
information. However, since the same information was available from each
source, this was not entirely surprising. The primary focus of the studies was
the actual behavior of the visitors within the gallery space.

The highlighting of links between paintings was not very successful in encouraging different patterns of movement between paintings. However, several

visitors were seen to look away from the painting they were standing in front of, to search for a linked painting identified in the audio content. This suggests that with stronger cues it might be possible to encourage different patterns of movement, in order to overcome visitors' tendency to walk around the perimeter of the room.

A particular highlight of the study was visitors' experiences with one particular painting that had a large number of content items associated with it. This painting contained a number of small details that were typically overlooked by visitors without the prompts of the PDA (and also by visitors with the printed guide). Spending more time in front of the painting prompted the delivery of information about more of the details and instructions on how to find them in the painting. One group of participants was observed to spend over three minutes (a long time to spend in front of one painting) pointing out details to each other and discussing what they were hearing. Several other visitor pairs were seen to engage in conversation about what they were seeing, conversation that was prompted by the information presented to them by the PDA guide. The system was thus successful in cuing informal collaborative learning for specific paintings.

CASE STUDY 3: MYARTSPACE

Background

The class trip to a museum or gallery is a long-established activity for many schools. Research suggests that, to maximize learning potential, such trips should ideally be preceded by pre-visit preparation for cognitive learning and orientation to the site to be visited. Such preparation "improves the chances of learning, especially if it involves integration of the school and museum learning, and provides opportunities for student involvement."[16] There is added value from post-visit activities that support the student to assimilate newly learned concepts, resolve possible misconceptions, and build on any increased interest and motivation that may have resulted from the visit.[17] One study suggests that "making the links between school and museum learning explicit, genuine, and continuous affords real opportunities for school students to have enjoyable learning experiences in both settings."[18] John Falk and Lynn Dierking suggest in chapter 2 that the in-museum, free-choice learning experience needs to be situated within the larger framework of the visitor's life, before and after the visit. Similarly, a school museum visit needs to be situated within the larger framework of school learning.

However, teachers and curators agree that, in general, they do not make best use of the learning opportunities offered by the class museum visit. At best there is a need to improve communication between museums, schools, and individual teachers, to encourage connected learning environments and tailored sessions.[19] At worst, the teachers do not plan the visit in advance, the children rush around the museum filling out worksheets, and their only lasting memory is of a fun but disorganized day away from school:

> On site, the traditional didactic display and the old-fashioned single-file tour around an exhibition accompanied by a worksheet ("death by worksheet") go hand-in-hand. In these circumstances, the pupils notice little except what the questions were about, and obtain most of the answers from each other or from a friendly attendant. Fortunately the days of the worksheet are numbered (I hope).[20]

Description

MyArtSpace is a mobile learning system that aims to address the discontinuity between museum and school learning by making a day out at the museum part of a sequence that includes setting a big question in the classroom, exploring it through a museum visit, reflecting on the visit back in the classroom or at home, and lastly presenting the results. The technology provides the essential link across the different settings.

The teacher starts by planning a class visit to the museum, consulting the MyArtSpace Teacher's Pack to prepare the trip. Typically, the teacher sets an open-ended question that students can answer by gathering evidence during the museum visit. For example, ten- to twelve-year-old students from a history class visited the D-Day Museum in Portsmouth, UK, which interprets the Allied landings during World War II. Their task was to collect evidence on whether D-Day was a triumph or a disaster for Britain.

At the museum, the students were each loaned a multimedia mobile phone, into which each student entered a personal identification code. A student can "collect" an exhibit by typing a two-letter code that is shown on a printed label beside the exhibit. This downloads relevant image and text information, and at the same time, the same image and text are sent to the student's personal Web space. The student is prompted to type in the reasons for "collecting" the exhibit, encouraging reflection upon the exhibit's relation to the big question at hand. After collecting an exhibit, the student is shown a list of others who have also collected it and prompted to find and talk with them face-to-face. In addition, the student can use the phone to create a personal interpretation of the visit by taking photos, recording audio, or writing text comments, all of which are sent to the website.

Back at school or at home, the students can view their collections (see figure 9.3). Each student's Web space shows a record of the visit, including the collected exhibits, plus captured pictures, sounds, and notes. They can also see and use items in the collections of other students in the class, as well as items provided by the teacher and the museum. The students can organize their collections into personal galleries (simple Web-based presentations) to present in

FIGURE 9.3

Screenshot of MyArtSpace on a PC, showing an image collected during a museum visit, one uploaded by the museum, and one uploaded by a teacher.

the classroom, or to share online. Access to the Web space is password protected, and the content published by the students is moderated to ensure privacy protection and appropriate use.

The primary use for mobile phones is usually making telephone calls and exchanging text messages. In MyArtSpace, these functions are disabled. Instead, the mobile phone is employed as a general-purpose computing device. The specially written software creates an environment in which students are able to collaboratively construct and share representations of their interpretations both within and outside school. Thus, the technology is used appropriately within the context of the learning task, regardless of the "expected" functionality.

Discussion

MyArtSpace was piloted between February 2006 and March 2007, during which time it was used by more than four thousand students, in three museums in the UK (the Urbis museum in Manchester and the Study Gallery in Poole, in addition to the D-Day museum). A series of evaluations were undertaken by the authors, including three full-scale trials with student groups.

We found that students using MyArtSpace in the museum engaged well with the learning task, "collecting" exhibits, examining and discussing their relevance, and annotating them with photos and sounds. A museum education officer described a typical school visit to the D-Day museum as lasting around twenty minutes, while students using MyArtSpace spent about one and a half hours exploring the venue, engaging with the exhibits. Teachers and museum educators agreed that the technology slowed students down by occupying them with activities that required their attention and kept them on task. The students' enthusiasm about using technology that is familiar (the phone) and "cool" (latest model) while on "school business" increased their motivation to engage. It is worth adding that, at the start of the project, the mobile phone used was a high-end model; by the end of the project, a year later, the phones were seen as out of date and had lost their "cool" status. Nonetheless, the mobile phone was still acceptable and retained the advantage of a familiar tool that young people related well to, eliciting positive reactions to the fact that it was a phone and not a typical museum guide device.

The technology proved an effective "conversational partner" that kept the students' attention on the exhibits, by providing them additional information in context, through exhibit collection; by enabling them to create their own representations through photographs and sounds; and by triggering discussions about the exhibits between students working in groups.

Not only did the technology enable students to access information in context, and even to create their own context through pictures, sounds, and text comments, but it also allowed students essentially to capture that context in the museum and have it transported back to the classroom. Everything they collected in the museum automatically ended up being part of a meaningful artifact (their personal collection) that they could take away from the museum and put to use in later classroom sessions.

The "durability" of the products created during a visit thus enables learning to cross locations and, as a consequence, to span over time. A student's personal online collection created on-site is available to them back at school or home after the visit. Moreover, the museum context itself extends into the classroom through the online "store" provided by the museum. In practice, this permits teachers to plan focused lessons that build up to and upon the visit. Although holding pre- and post-visit lessons is not standard practice for many teachers, the teachers in the trials were enthusiastic about the way MyArtSpace offers an easy route to planning such lessons and to making them enjoyable for the students. MyArtSpace thus enables learning to cross formal and informal contexts naturally.

The above summarizes the potential of the MyArtSpace technology to link contexts and to facilitate new communication channels. In practice, however, we sometimes observed students being disoriented when accessing the collected artifacts back in the classroom. In particular, some of the photographs and sounds collected in the museum did not make sense in the classroom. The context provided by the physical exhibit was lost when in the classroom, and children had not captured sufficient explanatory information along with the photo or sound to prompt their recall.

The service could be extended to less formal family and individual visits, with visitors registering for personal space on the MyArtSpace website and either being loaned phones or running the service on their own mobile phones. The latter would ensure that the visitor was familiar with the technology and would address the rapid obsolescence of hardware. However, using visitors' own phones would pose a significant technical challenge to develop software for multiple phone technologies, and it is unlikely that there could be a successful business model that would repay the development costs. A simpler service based on multimedia messaging or mobile e-mail may be more feasible for the future.

Thus, although the MyArtSpace phones enabled the rapid collection of information that was meaningful in the context of the museum visit, the collection process needs to mature, both technically and cognitively, if such information is always to remain meaningful within other contexts, such as the classroom. For example, improvements to the technology could allow students to directly link the photos, sounds, and written comments they record, together with the items they collect; currently, these are collected as separate pieces of data. In this way, the students' collected items could later appear within the context of the specific exhibits that prompted them. Cognitively, students could be trained to develop skills for capturing information in context and retaining important aspects of that context, for example, through meta-annotations. Overall, these suggested enhancements underscore the importance of context in the (re)construction of meaning, in this case as they are passed for the learner from one location to another.

CONCLUSION

This chapter has presented a theory of mobile learning that places learner mobility at the center of analysis. It has highlighted the opportunities that mobile technologies offer for learning in context, across contexts, and across time. It focused on the critical role of conversation and communication in mobile learning and the importance of engagement. The case studies have illustrated how the theory is played out in practice, and reinforced the central theoretical ideas, in the following ways.

The CAGE project as a context-sensitive application illustrates how technology can support learning by providing information related to learners' immediate context. Visitors move around a gallery and receive content at the right time and at the right level of detail. In addition to supporting the immediate learning situation (the painting being viewed), the information provided may propose, and prepare the learner for, new contexts (i.e., related paintings). Conversely, visitors may navigate through the information available by changing their location. This project highlights that context is not static; it is continually shifting as the learning experience changes.

The importance of collaboration and conversation in mobile learning is highlighted by the George Square system. Visitors utilized the technology to engage in a process of conversation and collaboration, which generated a shared understanding of the real and virtual space. This was achieved through spoken conversation between the visitors, using the shared resources provided by the technology. Discussion led to photographs being taken in the square and shared to provide further discussion, while the map made a common view of the space available to both visitors. Web pages provided access to information from experts, peers, and others, information that in turn led to discussion and promoted further exploration of the square.

The George Square system also demonstrated the utility of communication over time. The recommendation system offered information that previous visitors used when in a similar context, including places, Web pages, and photos taken. Similarly, current activity was stored for future visitors who may encounter a similar context. Like the CAGE system, these recommendations propose, and prepare the learner for, new contexts (i.e., places in the square and their digital representations).

The pre- and post-visit facility of the George Square system offered the visitor an integrated learning experience. The virtual visit may be made before, during, and after the physical visit. In particular, the post-visit allows more information to be obtained on interesting aspects of the visit, and discussions to take place with people not available during the visit itself, such as experts or teachers. Thus, the learning context of the visit—locations visited, photos taken, and Web pages browsed—is made available for later study at home or elsewhere.

The MyArtSpace project demonstrates how technology can support the integration of learning both within and across locations. Within the museum the students actively engage with the exhibits, taking photographs, recording audio, and capturing objects to create their own collections. The learning spans the transitions from school (pre-lesson), to the museum (visit), and back to school (post-lesson). The content that students collect at the museum is then available for them back at school. In addition, the museum context is transported into the classroom through the online museum "store." The students may also bring their learning into the home, where they can share their experience with family and friends.

During the museum visit, students discussed the exhibits, sometimes sharing photos they had taken and sounds they had collected. This use of shared media as stimulus for discussion mirrors the way the George Square visitors used photos and Web pages to facilitate discussion of their shared spaces (physical and digital).

MyArtSpace illustrated that the technology (mobile phone) was an effective conversational partner. It focused students' attention on the exhibits and extended the time they spent exploring the museum. Working in conjunction with fixed technology (PCs) in the classroom and at home, students' learning moved between different contexts. The technology utilized was the most appropriate to the task at hand, but the information and the facilities available were dictated by the learning need, not the capabilities of the device. For example, students could not make phone calls on the mobile phone because this facility was not appropriate to the task, while photos taken using the phone were automatically available on the classroom PC because this information was useful to continuing the learning task in the new context.

Together, these projects show the main aspects of technological support for mobile learning: (1) provision of information immediately relevant to the learner's current context (e.g., within CAGE); (2) the facilitation of conversations with others in order to discuss and share the current experience (e.g., the two visitors within the George Square project); and (3) the movement of context-related information over time and space (e.g., accessing in the classroom content collected at the museum).

THE FUTURE

The projects described in this chapter describe examples of the key factors that are needed to support mobile learning. Information made available must be in context, across contexts, and across time, along with support for conversation and communication. Mobile learning devices of the future will need to integrate all of these services to support effective mobile learning.

We see two areas of development as salient to fulfilling these requirements. First, most of us already carry technologies capable of supporting mobile learning. Mobile phones offer audio and textual communication and are usually also able to capture and share photos, sounds, and video, with GPS positioning now becoming common. The similarly ubiquitous MP3 player can deliver virtually unlimited speech and music—functionality also appearing on mobile phones. Indeed, mobile phones are capable computing devices in their own right, replacing the PDA while taking on other roles, such as MP3 player and digital camera.

Second, the continuing expansion of networking infrastructure is driving a move away from individual devices storing their own data. Already, e-mail may be accessed from anywhere; Web applications such as del.icio.us offer the same for bookmarks, Flickr for photos, and YouTube for video.[21] This move to remote storage will make mobile learning readily available on any device, not just the one being carried. Moving between devices will become increasingly seamless, allowing easy access to large screens, printers, powerful processors, and other devices.

For museums, these developments allow the visitors' experience to be expanded far beyond the individual artifacts. Pre-visits to a virtual representation of the museum (accessed on PC, TV, mobile device, etc., as convenient) will allow an overview to be gained before the visit, with questions and viewpoints to generate interest and motivation.

In the museum, a mobile device will be aware of the visitors' location, interests, previous learning, places visited, and people in their social network. The device will be able to integrate these to act as a personal guide, suggesting routes through the space. It will offer personally and immediately relevant information from the digital world—text, speech, sounds, pictures, and video—to illustrate the exhibits for the visitors' individual perspectives. This will work in conjunction with the museum's fixed displays: large screens will show pictures appropriate for the visitors currently present; fixed kiosks will set themselves according to the user's interests and preferences, including language and ability. As new technology becomes available, such as 3D displays, the experience will become increasingly immersive. On leaving the museum the digital visit will remain available, allowing further study and discussion of the experience.

The growing use of technology as an active medium for creating and sharing knowledge will allow the visit to go beyond the transfer of information from curator to visitor, allowing visitors to contribute their own thoughts and ideas, photos, and document recommendations. Future visitors will be offered such items from previous visitors with similar interests. Indeed, this principle may be expanded to create a new type of museum, curated by the people themselves, of which the "official" museum artifacts may be only a part. For example, a museum of transportation could be expanded by visitors (and nonvisitors) to include photos of modes of transport in use around the world.

Technologically, the future for mobile learning is upon us. Making that technology easily usable and applicable to the museum setting is now the major challenge.

ACKNOWLEDGMENTS

We should like to thank and acknowledge all those who assisted with and took part in evaluations for the three projects described here, and the contribution of nonauthor project partners to the work. MyArtSpace was funded by Culture Online, part of the UK's Department for Culture, Media and Sport. The service has now been commercialized as OOKL, see www.ookl.org.uk. The MOBIlearn project was supported by the EU 5th Framework Programme, see www.mobilearn.org. The George Square system was supported by the Equator EPSRC grant (GR/N15986/01), see www.equator.ac.uk.

NOTES

1. Peter Lonsdale, C. Baber, and Mike Sharples, "A Context Awareness Architecture for Facilitating Mobile Learning" (paper presented at Mlearn 2003—Learning with Mobile Devices, London, May 2003).

2. Gordon Pask, "Minds and Media in Education and Entertainment: Some Theoretical Comments Illustrated by the Design and Operation of a System for Exteriorizing and Manipulating Individual Theses," in *Progress in Cybernetics and Systems Research*, ed. R. Trappl and Gordon Pask, vol. 4 (Washington, D.C.: Hemisphere Publishing Corporation, 1975), 38–50; Diana Laurillard, *Rethinking University Teaching: A Framework for the Effective Use of Learning Technologies*, 2nd ed. (London: Routledge Falmer, 2002).

3. M. Smith, "Paulo Friere," in *Encyclopedia of Informal Education* (2005), at www.infed.org/thinkers/et-freir.htm (accessed August 29, 2007).

4. Giasemi Vavoula, "A Study of Mobile Learning Practices," Deliverable 4.4 for the MOBIlearn project (EU, IST-2001-37440; 2005), at www.mobilearn.org/download/results/public_deliverables/MOBIlearn_D4.4_Final.pdf (accessed August 29, 2007).

5. Wikipedia, home page, at http://wikipedia.org.

6. Google, "Google Mobile," at www.google.co.uk/gmm.

7. Flickr, "World Map," at www.flickr.com/map.

8. A. Ravenscroft, "Designing Argumentation for Conceptual Development," *Computers and Education* 34 (2000): 241–55.

9. Laurillard, *Rethinking University Teaching.*

10. Equator was a six-year Interdisciplinary Research Collaboration (IRC) beginning October 2000, supported by the UK Engineering and Physical Sciences Research Council, involving eight academic institutions.

11. Barry Brown, Matthew Chalmers, M. Bell, M. Hall, Ian MacColl, and Paul D. Rudman, "Sharing the Square: Collaborative Leisure in the City Streets," in *Proceedings of ECSCW 2005* (Paris: Spinger, 2005), 427–29.

12. Barry Brown and Matthew Chalmers, "Tourism and Mobile Technology," in *ECSCW 2003: Proceedings of the Eighth European Conference on Computer Supported Cooperative Work*, ed. K. Kuutti and E. H. Karsten (Dordrecht, Finland: Kluwer Academic Press, 2003), 335–55.

13. Giorgio da Bormida, Paul Lefrere, R. Vaccaro, and Mike Sharples, "The MOBIlearn Project: Exploring New Ways to Use Mobile Environments and Devices to Meet the Needs of Learners, Working by Themselves and with Others" (paper presented at the European Workshop on Mobile and Contextual Learning, Birmingham, UK, June 2002).

14. Peter Lonsdale, C. Baber, and Mike Sharples, "A Context Awareness Architecture for Facilitating Mobile Learning" (paper presented at Mlearn 2003—Learning with Mobile Devices, London, May 19–20, 2003).

15. J. Cross, S. Wooley, C. Baber and V. Gaffney, "Wearable Computing for Field Archaeology" (paper presented at the International Symposium on Wearable Computing Applications, Zurich, 2002).

16. Jeanette Griffin, "Research on Students and Museums: Looking More Closely at the Students in School Groups," *Science Education* 88, suppl. I (2004): S59–S70.

17. David Anderson, Keith B. Lucas, and Ian S. Ginns, "Theoretical Perspectives on Learning in an Informal Setting," *Journal of Research in Science Teaching* 40, no. 2 (2003): 177–99.

18. Griffin, "Research on Students and Museums," S67.

19. E. Johnsson, *Teachers' Ideas about Learning in Museums* (London: London Museums Hub, 2003).

20. G. Black, *The Engaging Museum* (London: Routledge, 2005), 168.

21. del.icio.us, "Social Bookmarking," at http://del.icio.us/; Flickr, home page, at www.flickr.com/; YouTube, "Broadcast Yourself," at http://uk.youtube.com/.

10

Interactive Adventures

Halina Gottlieb

The development of the next generation of personal museum technologies is a big challenge. The potential development directions—which include visitor tracking, personalization, multimedia, extending the visit, and harnessing visitors' own devices—appear infinite. This chapter explores a direction less taken by others but equally innovative: the packaging of traditional handheld technologies such as audio guides and PDAs into creative guides delivering "interactive adventures." Such guides generally build upon established paradigms but deviate from their predecessors in an effort to make the technology invisible to the user. Moreover, they often tailor content to specific visitor demographics and enhance visitor engagement by providing a perspective on content that better fits particular demographic groups. The projects mentioned here try to adhere to all of the above by replacing strictly technical devices with everyday things and toys. This chapter focuses on three projects that ran in three Swedish museum environments: Nationalmuseum (Stockholm), Universeum Science Discovery Center (Gothenburg), and Avesta Ironwork (Avesta).

The transition from traditional guides to interactive adventures adds a new influence in the development of handheld tools, one popularly referred to as interaction design. Interaction design seeks to establish a dialogue between products, people, and physical contexts, in order to anticipate how the use of content will affect comprehension and to determine an appropriate form.[1]

Hence, the theoretical framework used in this chapter comes from the educational and semiotic fields, within a sociocultural perspective.[2] It indicates a difference between those visitors who used the guides and those who did not, both in terms of their conduct in the exhibition space and their reflections of the museum environment. Broadly speaking, visitors who used a guide showed more interest and remembered more exhibition-specific facts than those who participated in traditional tours, conducted by a human guide.

STORYTELLING TOY ANIMALS

Introducing children to art in museums often aims at including the child in a discussion about art, in order to stimulate the child's own artistic interest and talent. The aim of the visit is to liberate "something" almost metaphysical, from inside the child. This method is very often based on the child participating in a group, and the learning is focused on these premises. What then about the individual young visitor? What can be done for her or him?[3]

The first two projects described in this chapter promote alternative ways to engage the young individual. Based on the established premise that engaging child participation and guidance is crucial to learning, these guides sought to incorporate greater levels of interactivity than more traditional guides, a feature essential to maintaining the interest of this visitor demographic.

The first project used stuffed animals as the physical packaging for an audio guide and was piloted at the Nationalmuseum in Stockholm. Two types of animals were available, each with their own characteristics: a bat, which had a cool, sleek appearance; and an owl, which was soft and cuddly. And beyond merely physical attributes, these characteristics were integrated into the style and voice of the guided tour. The guide was triggered by sensors at selected artworks and would tell a story related to the artwork when in proximity. The nature of the guided tour depended on the character chosen, as each animal had different stories to convey. Furthermore, each character could be customized by level of difficulty and language.

It was remarkable to observe how something as trivial as choosing an animal impacted a young visitor. Offering the child a choice of companion allowed him or her to quickly develop a personal connection to the guide and provided a sense of significance. For each work of art, the perceived uniqueness of the story and interactivity raised the child's attentiveness.

FIGURE 10.1

Audio guides often incorporate multiple layers of content, but there are no visual indications to show whether a companion is listening to the same audio or to something altogether different. Furthermore, it is presumptuous to assume that a young child will be able to navigate a complex structure of content. This added complexity may add the need for adult supervision, which could create a barrier between the child and the exhibition. But the visual identity of the stuffed animal allows the child to form a unique relationship with a personal tour at the exhibition.

This promotes interaction between visitors. If two children happen to stand in front of the same artwork with two different characters, their curiosity often compels them to compare stories. This reinforces learning of central themes and ideas, and spurs creative reflection. This vital component for learning seldom takes place with traditional handheld guides. With exactly the same content on their guides, unacquainted visitors have fewer reasons to strike up a conversation with each other. The knowledge that different guides bring different experiences to the tour may inspire children to revisit the exhibition with a different character. In some cases, it may even inspire a second visit to the museum.

Another issue is that some children are shy and introverted, and therefore may not benefit from traditional guides or children's group tours. The child may be afraid to venture out "alone" in the exhibition, and the impersonal nature of a traditional guide will lend no support. With a familiar toy and friendly character, however, the child can feel comfortable visiting the museum without joining a children's tour.

AN ANIMAL GUIDE FOR GIRLS AT A SCIENCE CENTER

The second example of an innovative guide for children was developed for an exhibition about the Swedish ecosystem at the Universeum Science Discovery Center in Gothenburg, called "The Water's Way." The audio guide allows the child to "listen" to animals living within the selected ecosystem, who narrate the story of their lives, telling of inherent conditions and impact of the natural environment. Visitors can choose from four animals—a wolf, beaver, adder, or salmon—all species that follow "The Water's Way" through Sweden.

Every audio guide consists of a headset and bracelet that indicates the type of animal the visitor has chosen. Moreover, each animal tells a different story, which is read by a different actor. Their audio is activated at stations throughout "The Water's Way" exhibit space.

Upon arrival at Universeum, children select their animal. The headband of each animal contains a receiver and a set of headphones. Eleven small identifiers located throughout the venue trigger the receivers to play the appropriate content when the child enters a predefined area. The next audio track comes on as soon as the child enters a different zone. This enables the child to do things at his or her own pace. A number of "senders," corresponding to the different stations, are situated at various places following "The Water's Way." These give the child different tasks to accomplish.

The concept of this guide is similar to the one at the Nationalmuseum. However, while the animal toys in the previous example provided a familiarity through which to mediate learning, the animals in this example provided an exotic, yet accessible, perspective on issues surrounding "The Water's Way" ecosystem. Presenting the perspective of the animals is not only more fun but also more down to earth and something that children can relate to.

Of the four animals, the wolf's audio guide is the longest and the most multimodal; it stops at most stations. It encourages the child to actively interact with the exhibit's environment and includes environmental sounds, sound effects, and dramatized music. The beaver's audio guide has no interaction but rather environmental sounds to enhance the feeling of being out in nature. The salmon and adder are both without any interaction and sound effects, consisting only of an actor's dramatization. These disparities allowed for the study of how increased levels of interaction and dramatization affect a child's experience of the exhibit and stories of the animals.

The success of this audio guide concept was demonstrated in a summative evaluation.[4] Most children agreed that the guides were too short and that they gladly would have participated in a longer tour. During the evaluation, participants were asked what they would have liked to see in an extended version of this project. A large proportion of their suggestions could be categorized as interactive, testifying to the popularity of such an approach. For instance, visitors wanted the animals to ask more questions, assign the children tasks (such as finding things in the milieu), solve puzzles, fetch things for the animals, and receive more multimedia content.

FIGURE 10.2

In addition to being a general description of an ecosystem, the guide also raised awareness of environmental issues. The unusual account of this from the perspective of animal characters provided an efficacious format for the dissemination of this important message to future generations. The personal and firsthand stories of the animal characters provided concrete and accessible examples, which young visitors could understand and relate to on an emotional level.

A FLASHLIGHT GUIDE AT AN INDUSTRIAL HERITAGE SITE

The final example of an innovative solution for a digital guide is an interactive adventure found at the old ironworks in Avesta. This region was one of the birthplaces of the Swedish steel industry. The disused blast furnace hall has been transformed into a dark, atmospheric exhibition of historical significance.[5]

This exhibition consists of an array of different multimedia installations, which illuminate such content as how steel was once produced here and what the work and living conditions were like for the workers. Some installations, for instance, provide a first-person account of the workers, while others feature a schematic animation illustrating the various steps in steel production.

Rather than rely upon traditional handheld devices, the exhibit designers chose to use a modified flashlight to serve as an interpretive guide. Special stations are demarcated by circular lamps; a visitor then activates the station by aiming the flashlight at the lamp while pressing the flashlight's button. When the installation receives the signal, the lamp switches off and runs the corresponding content.

Here we see how the traditional handheld guide can be replaced with an object that fits naturally into the exhibition's context. In addition to serving as an interpretive guide, the flashlight can also literally illuminate the dimly lit exhibition hall. A seemingly minor detail like this can have a huge impact on visitor experience. Investigating the innards of this old industrial building with a light beam provides the visitor with a sense of adventure and exploration.

Visitors can choose from two kinds of flashlights: the "educational flashlight" allows visitors access to information concerning the steel production process; the "narrative flashlight" provides a more evocative tour, describing daily life and the thoughts and aspirations of those who worked in this

FIGURE 10.3

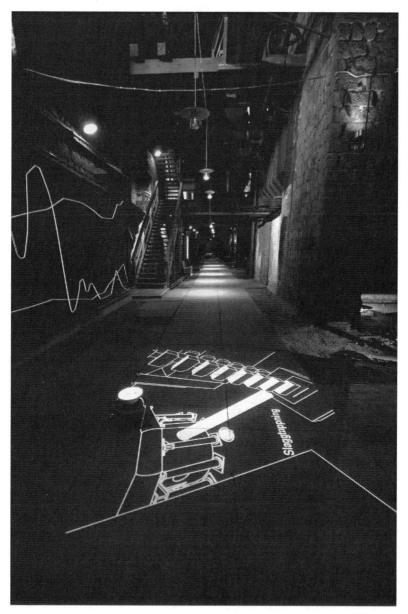

FIGURE 10.4

extreme environment. There is also built-in support for other flashlights, which might, for instance, feature other languages or levels of content. This allows the museum to cater to visitors with varying interests, enabling them to reach different visitor demographics, such as tourists and school groups, as well as the local population. Customizing the content for the different target groups is one way of improving visitor satisfaction.

CONCLUSION

Guides for visitors in the cultural heritage domain are going toward a mix of mobile, stationary, virtual or augmented reality and Internet technologies. This stirs up questions regarding what will constitute "good" interaction design in the future. The complexity of the design process, the unpredictability of usage, and the rapid development of technologies as design materials make concrete examples of good interaction the only stable foundation for future developers of handheld guides.

To develop innovative projects that resonate with visitors, technologies need to be implemented in novel ways to create engaging interfaces that will stimulate users. In the projects discussed here, the emphasis was on combining technologically innovative ideas with a design methodology and narrative. The objective was to produce creative results by a means referred to as the "collective designer"—an environment where it is possible to combine social ethics with innovative technology and design.[6]

All three projects were experiments in which traditional handheld technologies, in nontraditional guises, offered interactive adventures. In keeping with interaction design principles, the guides all featured minimal and intuitive interfaces, making most of the technology transparent to the user. A good way of attaining this interface simplicity was through camouflaging the technology behind familiar objects and symbols. (However, if such an object's behavior were to deviate too far from expected behavior, it could easily backfire and instead create an unwanted conceptual barrier.)

Hybrids between portable guides and immersive installations could offer interesting solutions for the interpretation of objects or themes. In the immersive category we include large-scale multimedia productions, often set designed, with or without user interaction. Multimedia shows (with digital and analog elements), "mixed media," or large-screen video formats can also provide an enveloping effect. This is also true of other forms of audiovisual productions that support or change spatial design with the help of light, projection, and sound. The third project discussed in this chapter, the ironwork in Avesta, is an example of such a hybrid production.

In immersive installations, the design is not evident but rather reliant upon a multitude of factors, such as purpose, the type of room, the choice of technology, combinations of media format, elements of user interaction, interplay with analog sources, and so on. In such spatially based productions the visitor is placed at the center of the plot, enveloped or engulfed by a certain atmosphere, a certain reproduction of a story; the technology constitutes a "digital set design." Often, several people can share the experience simultaneously, either by moving freely or by being guided at a certain pace. Portable guides could facilitate the exploration and navigation of such augmented, virtual, and immersive spaces.

In the 1990s, information technology researcher Mark Weiser coined the expression "ubiquitous computing" to denote a stage in the development of technology where it recedes into the background of our lives while still retaining its advantages.[7] It is paramount that the cultural heritage sector stands at the forefront of this development, to uphold its role in a postdigital age.

NOTES

1. Eric Stolterman, *Utskrifter av 20 intervjuer med systemutvecklare* (UMADP-WPIPCS 43.91). Umeå: Inst. för Informationsbehandling/ADB, Umeå Universite, 1991); P. Ehn, *Work-Oriented Design of Computer Artifacts* (Falköping, Sweden: Arbetslivscentrum: Almquist and Wiksell International, 1988).

2. James V. Wertsch, *Voices of the Mind: A Sociocultural Approach to Mediated Action* (Cambridge, UK: Cambridge University Press, 1991); R. Säljö, *Lärande i praktiken. Ett sociokulturellt perspektiv* (Stockholm: Prisma, 2000).

3. John Dewey, *The Later Works, 1925–1953*, vol. 10, *Art as Experience* (Carbondale: Southern Illinois University Press, 1987); George E. Hein, *Learning in the Museum* (London: Routledge, 1998); Eilean Hooper-Greenhill, *Museums and Their Visitors* (London: Routledge, 1994).

4. Halina Gottlieb, Helena Simonsson, S. Lindberg and L. Asplund, "Audio Guides in Disguise—Introducing Natural Science for Girls" (paper presented at Re-thinking Technology for Museums: Towards a New Understanding of People's Experience in Museums, Limerick, Ireland, June 2005).

5. The project was conducted by a group of artists and researchers from Smart Studio at the Interactive Institute and Avesta Cultural Services (Interactive Salon).

6. D. Schön, *Educating the Reflective Practitioner: Toward a New Design for Teaching and Learning in the Professions* (San Francisco: Jossey-Bass, 1987).

7. Mark Weiser, "The Computer for the 21st Century," *Scientific American* 265, no. 3 (1991): 73–76.

11

Afterword: The Future in Our Hands? Putting Potential into Practice

Ross Parry

With this book in our hand, we have here, in a way, our own *mobile guide*. Like the portable digital media it has attempted to describe, this volume has served, in one sense, as a "wayfinding device" through a new and largely undiscovered area of professional practice and new museology. The book has shown us ways we might want to navigate through these new areas of curatorship and design, visitor experience, and display. Equally, the volume you now hold has (it is hoped) served us as a helpful source of reference, a compendium of information into which we can delve and browse according to our own interests and expertise. Indeed, dense and portable, the very physicality of this book reminds us of how successful and rewarding, personal yet communal (not to mention historical and familiar) handheld media are—and how deft readers can be at using them. And yet, this collection of essays—this guide—leaves us with much to digest, and even more to ponder, on the past and future use of handheld mobile media within the museum. In particular, there is the not insignificant question of how all of this apparent potential might be realized in practice, for most museums—not in the world of high-financed research and development, or commercial-led innovation, but rather in the everyday, real world of curatorship.

SOMETHING OLD, SOMETHING NEW

First, it is evident that the idea of a handheld guide is, in itself, far from novel in the museum. The concept of a visitor either arriving with, or purchasing, or being presented with some kind of catalog, booklet, pamphlet, or leaflet to direct or substantiate the visit is, patently, not a modern invention. As long as there have been institutions and spaces that recognize themselves as museums or galleries, so there have been printed (we might say "mobile") material on offer to describe and expand visitors' engagement and enjoyment of those places.

Thomas Barber's eighteenth-century portrait of the housekeeper of Kedleston Hall (in Derbyshire in the UK) reminds us of how far back the conventions of guides and guiding go in the museum. According to Samuel Johnson (a visitor to the house in September 1777), it was Mrs. Garnett's duty as a "distinct articulator" to show visitors around the house and put into their hand a printed catalog of the fine artworks that were on display.[1] Barber's painting presents the benevolent Mrs. Garnett, guide in hand, ready to lead the viewer around the house's treasures. Two hundred years before the advent of portable digital catalogs, guides, and narratives, sites of cultural heritage were, it seems, already deploying handheld mobile media.

There is an equally long custom of empowering visitors with tools to build their own pathways, to pursue their own interests, or perhaps to choose the pace of their movement through a collection. Equally established is the practice of curators writing content for media that might move or leave the museum (books, catalogs, visitor guides, and trails), as much as writing content that will remain a fixture (such as a text panel or object label). There is certainly a tendency to enthuse about the innovation of handhelds in this book. But we might do well to keep in mind the established practices and protocols of writing for portable media, and providing guides to visitors, that have characterized and shaped the notion of "exhibition" and "visiting" for a large part of museums' long histories. The history of museum technologies helps us to not only understand where new technology has come from and why but also appreciate the cultures and contexts into which it might find itself. As ever with digital media in the museum, we are wise to look back as much as we look forward.

At first glance it seems that the novelty of digital handhelds rests less in the *concept* than in the *medium*. The authors here agree that, unlike the guides of previous centuries, the mobile digital media of today can accommodate substantial amounts of multimedia content. Equally important is the evidence that these guides support fast searching and cross-referencing that appear more responsive, current, or "intelligent" than the indexes and contents pages of the fixed printed word.

However, crucially, there are other times when the handheld ceases to be just a guide and when it aspires instead to be something else. And it is here that museums are presented with not just a different medium but also an entirely different concept. We see, for instance, the device performing a two-way communications role; this is the digital handheld as phone or messaging system. Elsewhere, we see the handheld as recording device—of images, audio, or text. In these instances it is both the medium *and* the concept that are inventive and new.

Understandably, confronted with an entirely different channel of communication with visitors and an unprecedented tool to enhance visitors' experience, it is these potentials in particular that demand the attention of curators, visitors, and cultural commentators alike. Even with one eye on the past and aware of longer traditions of curatorship, and even with a resistance to allowing technology to determine practice, it is nonetheless hard to ignore the questions posed by the presence of this novel sort of medium. We may not choose to use them, and we may not be in a position to use them, and yet their existence alone requires our attention at the very least. It might be not the specific instantiation of the technology that is the significant outcome but rather the catalytic effect of the technology on how the museum thinks and makes assumptions about its purpose and provision. History shows us that the anomalous arrival of new technology in museums can invariably (if not instantly) lead to a reprogramming of practice. In these chapters we glimpse how this reprogramming is playing out for digital handhelds.

SOMETHING EMPOWERING

If we step back a little from the detail and specificity of the various case studies and theses of this volume, we can begin to see some areas where patterns and broader consensus become discernible. First, what most authors in the volume agree upon is the usefulness of mobile handheld digital media in meeting the different needs of different learners. A museum will always be advantaged by being able to make use of a tool that affords yet another means of interaction; that appears to privilege a different set of intelligences or skills; that seems to carry with it a certain cultural resonance that might appeal more to some audience segments than to others; or that might be more sympathetic to certain types of expository space. Therefore, digital handhelds are just another (recent) addition to the curator's "media toolkit," another device, "draw," or "hook" for which to reach. Like so many other new technologies before them (digital and predigital), they simply add another hue, another tone to the curator's palette—to use or not to use.

Looking across the essays in this volume we begin to make out a much higher resolution picture both of the specific contexts and the specific types of digital handhelds that have shown to be effective. We see glimpses of what intelligences, what learning styles, what cultural groups, and what spaces digital handhelds have shown to be effective for. After all, visitors enter the museum with their preferred learning styles, in different visiting modes, in different social groups, and from different cultural backgrounds. They make meaning in different ways, have different types and levels of literacy, and carry different experiences and levels of subject knowledge. They have different expectations for that visit on that day in that museum, and they have different skills and attitudes toward different types of interpretive media. The exhibitions they move within are of differing sizes, temporary and permanent, set within different academic and conceptual frameworks, with different levels of curatorial authority and narrativity, containing objects of different dimensions and media, designed to convey different messages and offer different learning outcomes, with different opportunities for physical, social, and intellectual engagement. Furthermore, the contexts of all of these functions may vary across the size and type of organization, as well as across the different cultures of professional practice across the world.

Moreover, as we have already observed, the term digital handheld covers a range of functionalities from pocket PC to mobile phone, to camera, to recording device—each of which may carry different expectations of use and application, and each of which may carry different design, technical, infrastructural, and financial constraints and issues. And it is evident that there have already emerged a number of recognizable communicative modes and functions that these varied media might perform in the museum: from wayfinding to a sourcing device, from assistive technology to interpreter to storyteller. In other words, visitors, exhibitions, cultures, and technology can each be viewed across their own expansive horizons of possibilities. Our ongoing discussions and experiments, applications, and developments with handhelds should therefore be prepared to always confront (rather than simplify) this complexity. A growing understanding of their affordances, and a growing sophistication in the discourse of digital heritage, will both be demonstrated when this degree of subtlety and exactitude is routinely made in our discussions about handheld media. Encouragingly, the differentiations made in this volume would appear to set us on the right course.

Second, there is a clear sense in which digital handhelds equip the museum with new forms of assistive technology that can make a larger amount of the museum's provision more accessible to a wider visiting public. And yet, the studies presented here also alert us to how much more there is to explore and understand on how digital handhelds might be targeted to empower visitors who have normally been disabled by traditional gallery design. In this regard, museums still need to nuance and differentiate more sensitively between the subtle but crucial differences that can exist between different forms of impairment and disability, the self-identities that may or may not be associated with those impairments and disabilities, as well as the personal mitigating strategies that may or may not be already in use by the visitor.

Third, several of the chapters in this volume provide further evidence of the museum as a multichannel, decentered, and disaggregated service and experience. This stands in contrast to orthodox museography and traditions of the "visit event." We glimpse in this book the many ways in which both the locus and the digital connectivity of a museum experience are being relocated and rewired in the informational age. We see visitors beginning their museum experiences online and at home before they arrive at the physical site. We see visitors staying online as they explore the gallery. We see visitors resuming a museum experience and continuing their relationship with the museum (through connected media) outside and away from the physical institution.

There are, of course, antecedents and well-established protocols for these kind of protracted, atomized, and dispersed encounters with a museum. Museum publishing, community outreach, and inquiry services, not to mention marketing and loans services, have already embedded within curatorial practice and visitor expectation the principles of distance learning, broadcast, and diffusion. (It is a fallacy to see the reticulated and disseminating museum as a product exclusively of the Web.) Mobile media have, therefore, to a great extent, plugged into this existing infrastructure and behavior. And yet, their unique functionality has also enabled connections that are potentially more fluid, unplanned, targeted, and dialogic. Untethered media have allowed visitors to be untethered from not just specific narratives or interpretations within the exhibition space but also, potentially, the museum venue as a whole.

AND YET, SOMETHING UNRESOLVED

Equally striking in this volume, however, have been the points of tension. The essays here have highlighted three areas in particular where authors either agree on these points of tension or where opinion or practice has tended to diverge.

First, the question remains as to whether handhelds work toward or against social experiences in the museum. We see, for instance, applications that tend to focus on personalizing and customizing the museum experience for a single visitor. The emphasis in these applications is on giving the visitor greater control and choice, providing more private and individualized attention. Curiously, even though these applications would appear to plug into well-established traditions of private reflection and "self-directing and self-managing" visitors,[2] they are nonetheless criticized as undermining an important characteristic of museum visiting. This is especially ironic at a digital moment at which elsewhere (online) digital media have, conversely, become synonymous with greater connectivity, dialogue, sharing, and community building amongst museum Web users.[3] It might be argued that there is something inherent within digital media itself (the human-computer interaction) that works to privilege an individual user and a singularity of use. The criticism relates to the "the selfishness of the single monitor" and the "hyperindividualation of the subject."[4] However, it is equally evident from some of the other projects and applications described in this book that the same media (with the same casing and display) can in fact, in other circumstances, support high levels of collaboration, dialogue, and a social experience. In these instances the handheld media connect visitors to the institution or to other visitors. The future design and use of digital handhelds will continue, no doubt, to recognize this divergence and difference in application, between something personal, individual, and private and something collective, communal, and public. Helpfully, these chapters have demonstrated that both have their place; that they are not mutually exclusive; and that neither is perhaps without its critique.

The second key contention with this book would appear to relate to the relationship between digital handhelds and narrative. Specifically, there appears at times to be an anxiety over the loss of clarity that may result from the use of a personal guide or digital assistant—especially when that guide is providing "random access" content. The benefits of a visitor being able to use a handheld to "delve deeper," or make serendipitous personal connections across a collection and exhibit, are potentially offset by disruption that this idiosyncratic pathway might have on the story the museum is attempting to tell. Narratives are a defining part of museums. They have been embedded within the museum concept for as long as we have been able to identify something called a museum. Narrative is a common framework for interpretation—it is an ordered way to make sense of the world. It has been an enduring device by which to organize and explain museum content, particularly in terms of the range of subject disciplines within the museum. And through these disciplines and academic traditions, the museum continues to be literally full of dominant narratives. The paleontological narrative, for example, presents time through space, our past through the fossil record, and a stratification of time beneath our feet—piecing together the story—just as the evolutionary narrative presents a story of adaptation of life on earth. The art-historical narrative (of schools and periods of art) still directs the hang of many of our art exhibits. The historical narrative, from metanarratives of the Enlightenment, to teleological Whiggish approaches, to the more deterministic, directional approaches of the Marxist historiography, all still sit beneath so many of our exhibitions. Even within the more polyvalent historiography of poststructuralism, storytelling has a key part to play. The disciplines of academe that have been so influential on shaping the epistemological frameworks of the museum (indeed of which the museum was intended to be a mirror of) have had narrative at their core.

However, in another sense, narrative has had a *performative* role—as an enthralling way to convey meaning. Alongside their illustrative, instructional, and educational (sometimes soberly didactic) role, museums have also always aspired to entertain, engage, and delight. Theatricality and spectacle are as much part of museums' histories as are the disciplines of the academy. And with this, storytelling has continued to be a powerful device with which to structure this performance.

Third, narrative has had a *constructive* and practical role to play in museums. Narrative has been the structured way of directing visiting experience and flow. The linear sequence of a story has matched the sequential and temporal nature of the visit event. Moreover, the architectural shells of museum buildings (designed or inherited) have presupposed or determined a linear, sequential procession of exhibits and visiting. Therefore, in its interpretive, *performative*, and *constructive* function, narrative has (enduringly) been integral to the museum experience. Museums are themselves a form of narrative. Even when museums have knowingly shaken off didactic models of communication, storytelling and narrative have remained a powerful and crucial way of ensuring that the resulting approach to exhibition does not lose focus or slide into intellectual relativism.[5] The question, therefore, is whether the "embarrassment of riches" (to use the words from one chapter) presented by digital handhelds, and the impulse to serve a wider array of audiences and needs, might ultimately dilute or disrupt the engaging stories the museum is so well placed to tell. This book admits (although perhaps not always directly) to a potential tension that can exist between the clarity and selective, framed singularity of the exhibit and the seemingly bottomless, encyclopedic variety of the handheld guide.

Finally, a third important question that emerges from the chapters in this volume relates to visitors generating digital content in the museum. The ability of digital technology to not just disseminate but also receive digital content results in a new channel of communication with visitors in the museum, and with it a new dynamic in the relationship between institution and individual. However, inevitably, this new channel (and this new user-generated content) brings with it new curatorial questions—many of which are still to be fully answered in the context of digital handhelds. Now that museums have the technology to elicit multiple formats of feedback, comment, content, and intervention from their visitors (before, during, and after a visit event), they are perhaps struck with the simple question of whether this is something they really want to do. Almost as soon as the revolution in user-generated content was upon us, high-profile commentators in other parts of the media and cultural industries (such as television) had already begun to air their concerns about the overuse of audience participation and contribution. The debate for television (as it worked through its migration from an analog to digital platform and all the possibilities this new system might afford) was whether "we've got too interested in the *way* we deliver what we do, at the expense of *what* we deliver."[6] In other words, now the technology is here to do it, it would not be misplaced for museums to ask themselves a series of questions about *why* and *how* they would want to encourage user-generated digital content. Is the museum expecting too much from its visitors? Is a "passive" visit by a visitor necessarily a bad thing? How will a museum determine the value of user-generated content? How will it ensure user-generated content is useful and successful? And, in terms of rights management, information management, permissions, workflow, and strategy, are museums ready to deal with a future of user-generated content? This book has, at the very least, opened the door to these important future discussions.

AND SOMETHING UNSAID

Digital handhelds will need to negotiate a contemporary discourse of digital heritage that is currently dominated by use of the Web. Reading through various professional and governmental reports and "roadmaps" for the future of cultural heritage (and certainly digital cultural content) one could be forgiven for thinking that the prospect is exclusively "Web-shaped."[7] To take the example of the UK: in its first attempt to confront (and articulate a vision of engaging with) the digital media explosion of the 1990s, policy makers all too reflexively equated the "new museums" of the new digital and learning age with specifically Internet technologies—"Information and Communication Technology" (ICT) appeared almost exclusively to mean "the Web." The late-1990s vision of a "digital museum" did indeed include "interactive gallery exhibits, participatory activities, personal digital guides."[8] But these references were largely made in passing, and, instead, future development was, more squarely, seen in terms of "interactive websites and online services" and "online information to help in planning visits."[9] When, in 2001, Resource (then the UK's body responsible for museums, libraries, and archives) sought to design a strategic framework for the future of ICT and "provide a bridge between strategy and action," it did refer to the personalized interpretation and planning of visits "perhaps by mobile device—palmtop or mobile phone."[10] However, overwhelmingly, it was to the Web that the government looked to as a medium for its main priorities of service delivery and social development. The Web was the key, so it seemed, to "touch the lives of everyone."[11] Even by 2005, with the convergence and blending of media aplenty, "digitization" was more often associated with Web-based initiatives and projects.[12] In the decade around the new millennium, research and development in digital heritage was anchored to Web-based technologies and projects. This has by no means excluded handheld and mobile media, many of which have used these same technologies. However, more often than not, the focus of funders, policy makers, and technology developers was on the Internet and museums' online channels, rather than on the design of in-gallery interpretive media. The debates and applications of digital handhelds will, therefore, need to make themselves heard within a professional and academic discourse skewed in recent years to the Web.

There also remain a series of other practical barriers and blocks to the institutional adoption of digital handhelds. For instance, the history of museum computing shows museums to be risk averse when adopting new technology.[13] With the "state of the art" constantly on the move, museums continue to require a robust body of evidence to sanction an investment into neoteric forms of interpretation, display, and communication. This is particularly apposite to organizations that perceive a "digital project" as something that "can age very quickly."[14] Similarly, even today, digital media (including mobile media) battle against perceptions and connotations of "dumbing-down."[15] These are the concerns that with the adoption of media like digital handhelds museums might "turn themselves into high-tech, interactive theme parks" and put most of their collections in storage.[16]

Moreover, there are a series of other, very real practical barriers to the more widespread and wholesale adoption of digital handhelds. First, there are no recognized sector standards for the design and application of digital handhelds in museums. Museum documentation, the writing of interpretive strategies, the conduct of environmental auditing and preventative conservation, and even the writing of object labels have all, to some extent, standards that support their use. Standards allow us to be more efficient, build productively on existing expertise, reduce risk, work in an informed way, create opportunities for interoperability and collaboration between projects, and allow for systematic evaluation and benchmarking within and across provision.[17] Europe's "Lund principles" might provide hope for such standards to emerge for mobile media in the museum;[18] however, these reference points, frameworks, and guidelines are not yet in place. Indeed, more generally the museum computing community still looks for more standards in the production and management of digital cultural content. It is still the case that to facilitate the exchange of information, museums currently can only draw upon "a number of weakly related archiving standards."[19]

The novelty and innovation of digital handhelds brings with it the needs for new skill sets within the museum profession. Since the moment computers arrived in the museum in a visible and systematic way in the late 1960s (almost exclusively in an information management role), there has been a call for a concurrent up-skilling of the workforce.[20] Subsequently, with the advent of the Web, governments and the museum sector were quick to identify the need for "new skills to create, manage and maintain participatory and truly interactive digital applications."[21] And yet, the skills related to the use and development of mobile media remain in a state of nascency for the museum. For instance, our schemes and techniques for writing text in the gallery have reached a state of maturity and sophistication,[22] and practitioners are aware of the new skills required for "writing for the Web." However, museums and training providers alike are only beginning to recognize the other skills that might be required for "writing for mobile media."

There is little said in the book about the initial cost and the cost of maintenance, upgrade, and support of the technologies described. When looking to make creative steps with digital media, many museums have historically been (and, indeed, remain) hamstrung by financial constraints.[23] Existing revenue budgets are still a barrier to many institutions plotting a development program that might include a substantial new investment in digital interpretive media.[24] Behind the commercial hype of flagship projects and specifically funded research and development initiatives, there exists, for most museums, a distinct lack of capacity that still restricts "their ability to use their collections to provide opportunities for learning, inspiration and enjoyment."[25] Some of the innovation, research, and development described in this book (which is laudable) is, in actuality, beyond the capacity of many institutions. A number of the projects trialed in this volume are with large European and North American organizations. In contrast, the stark reality is that less than 10 percent of all cultural heritage institutions in Europe are estimated to be in the position to participate in the digital era:

> The big majority of memory institutions—the local museum focusing on the history of a village, the community or church library or the highly specialised historic archive—do not even possess the human, financial and technological resources to accomplish the most basic things, such as digitally cataloguing their holdings or establishing a web presence.[26]

Despite many years of investment, access to ICT is still restricted across museums—one survey in the UK in 2002–2003 concluded that access was poor, and 15 percent of nonnational museums had no access at all. Other sites wanted it but could not afford it.[27] When digital media is a core function and embedded in the institutional mission, function, brand, and provision of the museum, the capacity, resources, and skills to support innovations with digital handhelds will be prevalent. Until then, the sector and the technology providers will need to continue to negotiate these barriers. This will mean learning to make a persuasive, institution-centered, visitor-orientated, evidenced case for digital handhelds.

IS THE FUTURE IN OUR HANDS?

We have reached the end of this book, and a few things seem particularly clear. First, that "guides" and guiding are (and to a great extent have always been) an innate element of museum exhibitions. There is no novelty or anomaly in visitors exploring exhibits supported by a map, a catalog, or a booklet. Even the first-generation guides of the electronic age (such as the audio tour) have become a mainstay of many institutions and the staple companion for many visitors. Second, it is apparent that the majority of visitors are walking through the doors of the museum with sophisticated handheld media of their own already in their pocket. Just like pulling out a pair of spectacles to read a label, checking a watch to plan a rendezvous, or reaching for a jotter to note down an interesting fact or make a sketch, so today the "tools for visiting," the paraphernalia that visitors (that we all) carry to navigate our way through the everyday, is becoming routinely joined by the mobile phone, the digital camera, and the music player—if not all in one device. Third, today's museography emphasizes a social, multisensory, responsive, object-based, visitor-centered experience. The emergent orthodoxy is of museums and visitors both expecting visits to be active (whether physically, intellectually, or socially) and expecting visitors to be valued—their needs, their perspectives, and their abilities.

We are left, then, with an exciting vision of convergence. This is a convergence between the tradition of guiding, the tools of visiting, and the priorities for curatorship. Compellingly, the projects described in this volume locate digital handhelds at the confluence of these practices. Unlike some of the more incongruous computer installations in the museum's past, digital handhelds therefore have the potential to slot into practices already present. The presence of these devices (or, at least, visitors' increasing familiarity and routine experience with these devices) is altering the assumptions the museum makes about the potentials of visiting. Mobile media and handhelds—connected, locative, ambient, and personalized—have emerged as both agents and epitomes of the modern museum. This is part of the vision of the Information Society, "where the emphasis is on greater user-friendliness, more efficient services support, user-empowerment, and support for human interactions."[28] But this alone does not ensure handhelds' future use in the museum. As has always been the case with new media in museums, the future of digital handhelds will not depend on potential alone. The practice and research around handhelds will need to prove that they are reliable and affordable, that the institution has the skills to manage them, and that the technology represents a long-term shift in exhibition media and visiting practice, rather than being a modish, voguish, and transitory fad. Moreover, the adoption of handhelds will also rest upon whether they capture (rather than contort), complement (rather than detract), and match (rather than complicate) both curators' and visitors' expectations of a museum experience. History seems to show that (successful) new museum media always need to harmonize with existing practice.

NOTES

1. A. Laing, *In Trust for the Nation* (London: National Trust, 1995.)

2. Tony Bennett, "Speaking to the Eyes: Museums' Legibility and the Social Order," in *The Politics of Display: Museums, Science, Culture,* ed. S. Macdonald (London: Routledge, 1998), 31.

3. Susan Chun, R. Cherry, D. Hiwiller, Jennifer Trant, and Bruce Wyman, "Steve.museum: An Ongoing Experiment in Social Tagging, Folksonomy, and Museums," in *Museums and the Web 2006: Proceedings,* ed. Jennifer Trant and David Bearman (Toronto: Archives and Museum Informatics, 2006), at www.archimuse.com/mw2006/papers/wyman/wyman.html (accessed May 25, 2007); S. Chan, "Tagging and Searching: Serendipity and Museum Collection Databases," in *Museums and the Web 2007: Proceedings,* ed. Jennifer Trant and David Bearman (Toronto: Archives and Museum Informatics, 2007), at www.archimuse.com/mw2007/papers/chan/chan.html (accessed May 25, 2007).

4. S. Cubitt, *Digital Aesthetics* (London: Sage, 1998).

5. M. O'Neill, "Interpretation Philosophy behind Kelvingrove" (Glasgow Museums, 2006), at www.glasgow.gov.uk/en/Visitors/MuseumsGalleries/KelvingroveMediaOffice/NewKelvingrovePhilosophy/Kelvingrovephilosophyinterpretation.htm (accessed November 1, 2007); Peter Higgins, "The Design Zone," *Museum Practice*, no. 22 (2003): 33.

6. Jeremy Paxman, "What's Wrong with TV?" James MacTaggart Memorial lecture, MediaGuardian Edinburgh International Television Festival, August 2007, at www.telegraph.co.uk/news/main.jhtml?xml=/news/2007/08/25/npaxspeech125.xml&page=3 (accessed November 1, 2007).

7. DigiCULT, *The DigiCULT Report: Technological Landscapes for Tomorrow's Cultural Economy—Unlocking the Value of Cultural Heritage* (Luxembourg: Office for Official Publications of the European Commission, 2002).

8. National Museums Directors' Conference, *A Netful of Jewels: New Museums in the Learning Age* (London: Author, 1999), 13.

9. National Museums Directors' Conference, *A Netful of Jewels*, 5.

10. Resource, *Information and Communications Technology and the Development of Museums, Archives and Libraries: A Strategic Plan for Action*, draft for consultation (London: Author, 2001), 1, 5.

11. Resource, *Information and Communications Technology*, 4.

12. Scottish Museums Council, *Museums, Galleries and Digitisation: Current Best Practice and Recommendations on Measuring Impact* (Edinburgh: Author, 2005).

13. Ross Parry, *Recoding the Museum: Digital Heritage and the Technologies of Change* (London: Routledge, 2007), 117–31.

14. Scottish Museums Council, *Museums, Galleries and Digitisation*, 50.

15. Parry, *Recoding the Museum*, 61–62.

16. S. Spalding, *The Poetic Museum: Reviving Historic Collections* (Munich, Germany: Prestel, 2002), 9.

17. David Bearman, "Standards for Networked Cultural Heritage," *Archives and Museum Informatics* 9, no. 3 (1995): 279–307.

18. DigiCULT, *The DigiCULT Report*, 42–43.

19. DigiCULT, *The DigiCULT Report*, 263.

20. Parry, *Recoding the Museum*, 123–24.

21. National Museums Directors' Conference, *A Netful of Jewels*, 17; Resource, *Information and Communications Technology*, 8.

22. L. Ravelli, *Museum Texts: Communication Frameworks* (London: Routledge, 2006); A. Grey, T. Gardom, and C. Booth, *Saying It Differently: A Handbook for Museums Refreshing Their Display* (London: London Museums Hub, 2006), 62–71.

23. Parry, *Recoding the Museum*, 120–23.

24. Scottish Museums Council, *Museums, Galleries and Digitisation*, 51.

25. Resource, *Renaissance in the Regions: A New Vision for England's Museums* (London: Regional Museums Task Force for Resource, the Council for Museums, Libraries and Archives, 2001), 18.

26. Clore Duffield Foundation, *Space for Learning: A Handbook for Education Spaces in Museums, Heritage Sites and Discovery Centers* (Arts Council England, Arts Council of Northern Ireland, Scottish Arts Council, Clore Duffield Foundation, Cabe Education, Department for Education and Skills, Design Commission for Wales DCMS, Heritage Lottery Fund, MLA, 2004), 256.

27. Clore Duffield Foundation, *Space for Learning*.

28. K. Ducatel, M. Bogdanowicz, F. Scapolo, J. Leijten, and J-C Burgelman, *ISTAG: Scenarios for Ambient Intelligence in 2010* (Seville, Spain: European Commission Community Research, IPTS, 2001), 1.

Bibliography

Adelman, L. M., L. D. Dierking, K. Haley Goldman, D. Coulson, J. H. Falk, and M. Adams. *Baseline Impact Study: Disney's Animal Kingdom Conservation Station.* Technical report. Annapolis, Md.: Institute for Learning Innovation, 2001.

Adelman, L. M., J. H. Falk, and S. James. "Assessing the National Aquarium in Baltimore's Impact on Visitor's Conservation Knowledge, Attitudes and Behaviors." *Curator* 43, no. 1 (2000): 33–62.

Allen, Sue. *Finding Significance.* San Francisco: Exploratorium, 2004.

———. "Looking for Learning in Visitor Talk: A Methodological Exploration." In *Learning Conversations in Museums*, edited by Gaea Leinhardt, Kevin Crowley, and Karen Knutson, 259–303. Mahwah, N.J.: Erlbaum, 2002.

Allen, Sue, and Joshua Gutwill. "Designing with Multiple Interactives: Five Common Pitfalls." *Curator* 47, no. 2 (2004): 199–212.

American Association of Museums. *Everyone's welcome.* Washington, D.C.: American Association of Museums, 1998.

Anderson, Chris. *The Long Tail: How Endless Choice Is Creating Unlimited Demand.* London: Random House Business Books, 2006.

Anderson, David. "Understanding the Impact of Post-Visit Activities on Students' Knowledge Construction of Electricity and Magnetism as a Result of a Visit to an Interactive Science Centre." Unpublished PhD diss., Queensland University of Technology, Brisbane, Australia, 1999.

Anderson, David, and Keith B. Lucas. "The Effectiveness of Orienting Students to the Physical Features of a Science Museum Prior to Visitation." *Journal of Research in Science Teaching* 27, no. 4 (1997): 485–95.

Anderson, David, Keith B. Lucas, and Ian S. Ginns. "Theoretical Perspectives on Learning in an Informal Setting." *Journal of Research in Science Teaching* 40, no. 2 (2003): 177–99.

Anderson, David, Keith B. Lucas, Ian S. Ginns, and Lynn D. Dierking. "Development of Knowledge about Electricity and Magnetism during a Visit to a Science Museum and Related Post-Visit Activities." *Science Education* 84, no. 5 (2000): 658–79.

Aoki, Paul M., Rebecca E. Grinter, Amy Hurst, Margaret H. Szymanski, James D. Thornton, and Allison Woodruff. "Sotto Voce: Exploring the Interplay of Conversation and Mobile Audio Space." In *Proceedings of the SIGCHI Conference on Human Factors in Computing Systems: Changing Our World, Changing Ourselves*, 431–38. New York: ACM Press, 2002.

Archibald, Robert R. "Touching on the Past." Paper presented at the Social Affordances of Objects seminar, London, November 2006.

Ash, Doris. "Negotiations of Thematic Conversations about Biology." In *Learning Conversations in Museums*, edited by Gaea Leinhardt, Kevin Crowley, and Karen Knutson, 357–400. Mahwah, N.J.: Erlbaum, 2002.

Astor-Jack, T., K. K. Whaley, Lynn D. Dierking, D. Perry, and C. Garibay. "Understanding the Complexities of Socially Mediated Learning." In *In Principle, In Practice: Museums as Learning Institutions*, edited by John H. Falk, Lynn D. Dierking, and S. Foutz. Linthicum, Md.: AltaMira, 2007.

Bal, M. *Double Exposures*. London: Routledge, 1996.

Baldessari, John. "TV (1) Is Like a Pencil and (2) Won't Bite Your Leg." In *The New Television: A Public/Private Art*, edited by Douglas Davis and Allison Simmons, 22–23. Cambridge, Mass.: MIT Press, 1977.

Ballantyne, R., and J. Packer. "Solitary vs. Shared: Exploring the Social Dimension of Museum Learning." *Curator* 48 (2005): 177–92.

Bandelli, Andrea. "Talking Together: Supporting Citizen Debates." In *Many Voices: The Multivocal Museum*, at http://clearingatkings.com/content/1/c6/02/37/03/Bandellipaper.pdf (accessed November 5, 2007).

Baptista, Luciana. "BBC Collect: Stapler Trial at London Zoo: Preliminary Findings." Unpublished evaluation report for the Zoological Society of London, November 2005.

Barry, Alisa. "Creating a Virtuous Circle between a Museum's On-line and Physical Spaces." In *Museums and the Web 2006: Proceedings*, edited by Jennifer Trant and David Bearman. Toronto: Archives and Museum Informatics, 2006. www .archimuse.com/mw2006/papers/barry/barry.html (accessed June 26, 2008).

Bearman, David. "Standards for Networked Cultural Heritage." *Archives and Museum Informatics* 9, no. 3 (1995): 279–307.

Beazley, Ingrid. "Spectacular Success of Web Based, Interactive Learning." In *Museums and the Web 2007: Proceedings*, edited by Jennifer Trant and David Bearman. Toronto: Archives and Museum Informatics, 2007. www.archimuse .com/mw2007/abstracts/prg_325001001.html (accessed November 5, 2007).

Beer, Valerie. "Great Expectations: Do Museums Know What Visitors Are Doing?" *Curator* 30, no. 3 (1987): 206–15.

Bellotti, F., R. Berta, A. de Gloria, and M. Margarone. "User Testing a Hypermedia Tour Guide." *Pervasive Computing* 1, no. 2 (2002): 33–41.

Bem, D. J. "Self-Perception Theory." In *Advances in Experimental Social Psychology*, edited by L. Berkowitz, 1–62. Vol. 6. New York: Academic Press, 1972.

Bennett, Tony. *The Birth of the Museum: History, Theory, Politics.* London: Routledge, 1995.

———. "Speaking to the Eyes: Museums' Legibility and the Social Order." In *The Politics of Display: Museums, Science, Culture*, edited by S. Macdonald, 25–35. London: Routledge, 1998.

Bielick, S., and D. Karns. *Still Thinking about Thinking: A 1997 Telephone Follow-up Study of Visitors to the Think Tank Exhibition at the National Zoological Park.* Washington, D.C.: Institutional Studies Office, Smithsonian Institution, 1998.

Bitgood, Steve. "What Do We Know about School Field Trips?" In *What Research Says about Learning in Science Museums*, 12–16. Vol. 2. Washington, D.C.: Association of Science-Technology Centers, 1993.

Bitgood, Steve, and D. Patterson. "Principles of Exhibit Design." *Visitor Behavior* 2, no. 1 (1995): 4–6.

Bitgood, Steve, Beverly Serrell, and D. Thompson. "The Impact of Informal Education on Visitors to Museums." In *Informal Science Learning: What Research Says about Television, Science Museums, and Community-Based Projects*, edited by V. Crane, 61–106. Dedham, Mass.: Research Communications, 1994.

Black, G. *The Engaging Museum*. London: Routledge, 2005.

Bormida, Giorgio da, Paul Lefrere, R. Vaccaro, and Mike Sharples. "The MOBIlearn Project: Exploring New Ways to Use Mobile Environments and Devices to Meet the Needs of Learners, Working by Themselves and with Others." Paper presented at the European Workshop on Mobile and Contextual Learning, Birmingham, UK, June 2002.

Borun, Minda, Matthew Chambers, J. Dritsas, and J. Johnson. "Enhancing Family Learning through Exhibits." *Curator* 40, no. 4 (1997): 279–95.

Bourdieu, Pierre. *Distinction: A Social Critique of the Judgement of Taste*. Cambridge, Mass.: MIT Press, 1984.

Bowen, Jonathan P., and Silvia Filippini-Fantoni. "Personalization and the Web from a Museum Perspective." In *Museums and the Web 2004: Selected Papers from an International Conference*, edited by David Bearman and Jennifer Trant, 63–78. Toronto: Archives and Museum Informatics, 2004.

Bransford, J. D., A. L. Brown, and R. Cocking, eds. *How People Learn: Brain, Mind, Experience, and School*. Washington, D.C.: National Research Council, 1999.

Brown, Barry, and Matthew Chalmers. "Tourism and Mobile Technology." In *ECSCW 2003: Proceedings of the Eighth European Conference on Computer Supported Cooperative Work*, edited by K. Kuutti and E. H. Karsten, 335–55. Dordrecht, Finland: Kluwer Academic Press, 2003.

Brown, Barry, Matthew Chalmers, M. Bell, M. Hall, Ian MacColl, and Paul D. Rudman. "Sharing the Square: Collaborative Leisure in the City Streets." In *Proceedings of ECSCW 2005*, 427–29. Paris: Spinger, 2005.

Brown, R., and J. Kulick. "Flashbulb Memories." *Cognition* 5 (1997): 73–79.

Burch, Alexandra, and Benjamin M. Gammon. "The Museum as Social Space: Scaffolding the Scaffolder." 2006. www.kcl.ac.uk/content/1/c6/02/36/86/burch.pdf (accessed November 5, 2007).

Bush, Vannevar. "As We May Think." *Atlantic Monthly*, July 1945.

Cabrera, J. S., H. M. Frutos, A. G. Stoica, N. Avouris, Y. Dimitriadis, G. Fiotakis, and K. D. Liveri. "Mystery in the Museum: Collaborative Learning Activities Using Handheld Devices." In *Proceedings of MobileHCI 05*, 315–18. New York: ACM Press, 2005.

Cassels, Richard. "Learning Styles." In *Developing Museum Exhibitions for Lifelong Learning*, edited by Gail Durbin, 38–45. London: Stationery Office on behalf of the Group for Education in Museums, 1996.

Ceci, S. J. *On Intelligence: A Bioecological Treatise on Intellectual Development*. Boston: Harvard University Press, 1996.

Ceci, S. J., and U. Bronfenbrenner. "Don't Forget to Take the Cupcakes out of the Oven: Strategic Time Monitoring, Prospective Memory, and Context." *Child Development* 56 (1985): 175–90.

Chan, S. "Tagging and Searching: Serendipity and Museum Collection Databases." In *Museums and the Web 2007: Proceedings*, edited by Jennifer Trant and David Bearman. Toronto: Archives and Museum Informatics, 2007. www.archimuse .com/mw2007/papers/chan/chan.html (accessed May 25, 2007).

Chin, E. "What Have We Learned from the *Star Wars* Multimedia Tour?" Paper presented at the ASTC 2006 Annual Conference, Louisville, Ky., October 2006.

Chin, E., and C. Reich, with contributions of A. Gaffney. "Lessons from the Museum of Science's First Multimedia Handheld Tour: The *Star Wars: Where Science Meets Imagination* Multimedia Tour." Unpublished evaluation report, 2006.

Cho, Dana. "Grasping the User Experience." Paper presented at Collaborative Artifacts Interactive Furniture, Château-d'Oex, Switzerland, June 2005.

Chris, Cynthia. "Video Art: Stayin' Alive." *Afterimage* (March 2000). http://find articles.com/p/articles/mi_m2479/is_5_27/ai_61535391/pg_1 (accessed August 6, 2007).

Chun, Susan, R. Cherry, D. Hiwiller, Jennifer Trant, and Bruce Wyman. "Steve.museum: An Ongoing Experiment in Social Tagging, Folksonomy, and Museums." In *Museums and the Web 2006: Proceedings*, edited by Jennifer Trant and David Bearman. Toronto: Archives and Museum Informatics, 2006. www.archimuse.com/mw2006/papers/wyman/wyman.html (accessed May 25, 2007).

Clark, A. *Being There: Putting Brain, Body, and World Together Again.* Cambridge, Mass.: MIT Press, 1997.

Clore Duffield Foundation. *Space for Learning: A Handbook for Education Spaces in Museums, Heritage Sites and Discovery Centers.* Arts Council England, Arts Council of Northern Ireland, Scottish Arts Council, Clore Duffield Foundation, Cabe Education, Department for Education and Skills, Design Commission for Wales DCMS, Heritage Lottery Fund, MLA, 2004.

Coe, J. "Design and Perception: Making the Zoo Experience Real." *Zoo Biology* 4 (1985): 197–208.

Cognitive Applications. *The Micro Gallery: A Survey of Visitors.* Evaluation report for the National Gallery, London, December 1992. www2.cogapp.com/homeimg/Micro%20Gallery%20Survey%20Report.pdf (accessed December 21, 2007).

Cone, C., and K. Kendall. "Space, Time, and Family Interactions: Visitor Behavior at the Science Museum of Minnesota." *Curator* 21, no. 3 (1978): 245–58.

Crane, V., ed. In *Informal Science Learning: What Research Says about Television, Science Museums, and Community-Based Projects.* Dedham, Mass.: Research Communications, 1994.

Creative Research. "Evaluation of *Future Foods?*" Unpublished evaluation report produced for the Science Museum, London, March 1998.

———. "Genetic Choices? Results of a Visitor Evaluation." Unpublished evaluation report produced for the Science Museum, London, April 1997.

———. *Summative Evaluation of the British Galleries.* Evaluation report produced for the Victoria and Albert Museum, London, September 2002. www.vam.ac.uk/files/file_upload/5874_file.pdf (accessed December 21, 2007).

Cross, J., S. Wooley, C. Baber, and V. Gaffney. "Wearable Computing for Field Archaeology." Paper presented at the International Symposium on Wearable Computing Applications, Zurich, 2002.

Crowley, Kevin, and Maureen Callanan. "Describing and Supporting Collaborative Scientific Thinking in Parent-Child Interactions." *Journal of Museum Education* 23, no. 1 (1998): 12–17.

Csikszentmihalyi, Mihaly. *Flow: The Psychology of Optimal Experience.* New York: Harper Perennial, 1990.

Csikszentmihalyi, Mihaly, and K. Hermanson. "Intrinsic Motivation in Museums: Why Does One Want to Learn?" In *Public Institutions for Personal Learning*, edited by John H. Falk and Lynn D. Dierking, 67–78. Washington, D.C.: American Association of Museums, 1995.

Cubitt, S. *Digital Aesthetics*. London: Sage, 1998.

Cupchik, G. C., L. Shereck, and S. Spiegel. "The Effects of Textual Information on Artistic Communication." *Visual Arts Research* 20 (1994): 62–78.

Davenport, Thomas H., and John C. Beck. *The Attention Economy: Understanding the New Currency of Business*. Cambridge, Mass.: Harvard Business School Press, 2001.

del.icio.us. "Social Bookmarking." http://del.icio.us.

Denning, Peter J. "The Profession of IT: Infoglut." *Communications of the ACM* 49, no. 7 (July 2006): 16–19. http://cs.gmu.edu/cne/pjd/PUBS/CACMcols/ cacmJul06.pdf (accessed June 26, 2008).

Dewey, John. *The Later Works, 1925–1953*. Vol. 10, *Art as Experience*. Carbondale, Ill.: Southern Illinois University Press, 1987.

Dierking, Lynn D., and John H. Falk. "Audience and Accessibility." In *The Virtual and the Real: Uses of Multimedia in Museums*, edited by S. Thomas and A. Mintz. Washington, D.C.: Technical Information Services, American Association of Museums, 1998.

Dierking, Lynn D., and W. Pollock. *Questioning Assumptions: An Introduction to Front-End Studies*. Washington, D.C.: Association of Science Technology Centers, 1998.

DigiCULT. *The DigiCULT Report: Technological Landscapes for Tomorrow's Cultural Economy—Unlocking the Value of Cultural Heritage*. Luxembourg: Office for Official Publications of the European Commission, 2002.

Dillenbourg, Pierre. "Scripted Collaboration and Locative Media." Paper presented at the Swiss Summit on ICT: New Generation Networks, Fribourg, Switzerland, October 2005.

Doering, Zahava D. "Strangers, Guests or Clients? Visitor Experiences in Museums." *Curator* 42, no. 2 (1999): 74–87.

Doering, Zahava D., and Andrew J. Pekarik. "Questioning the Entrance Narrative." *Journal of Museum Education* 21, no. 3 (1996): 20–25.

Doering, Zahava D., Andrew J. Pekarik, and Audrey E. Kindlon. "Exhibitions and Expectations: The Case of 'Degenerate Art.'" *Curator* 40, no. 2 (1997): 126–41.

Ducatel, K., M. Bogdanowicz, F. Scapolo, J. Leijten, and J-C Burgelman. *ISTAG: Scenarios for Ambient Intelligence in 2010*. Seville, Spain: European Commission Community Research, IPTS, 2001.

Dulwich Picture Gallery. "E-Learning, Digit: A Technical Revolution at Dulwich Picture Gallery." www.dulwichpicturegallery.org.uk/sackler/elearning.aspx (accessed June 26, 2008).

Durbin, Gail. "Is There Anyone Out There? Finding Out About How Our Web Sites Are Used." In *Museums and the Web 2006: Proceedings*, edited by Jennifer Trant and David Bearman. Toronto: Archives and Museum Informatics, 2006. www.archimuse.com/mw2006/abstracts/prg_295000735.html (accessed November 5, 2007).

Eberbach, Catherine, and Kevin Crowley. "From Living to Virtual: Learning from Museum Objects." *Curator* 48, no. 3 (2005): 317–38.

"Editorial." *Museums Journal* 5, no. 60 (August 1960).

Ehn, P. *Work-Oriented Design of Computer Artifacts*. Falköping, Sweden: Arbetslivscentrum, Almquist and Wiksell International, 1988.

Ellenbogen, Kirsten M. "From Dioramas to the Dinner Table: An Ethnographic Case Study of the Role of Science Museums in Family Life." Unpublished PhD diss., Vanderbilt University, Nashville, 2003.

———. "Museums in Family Life: An Ethnographic Case Study." In *Learning Conversations in Museums*, edited by Gaea Leinhardt, Kevin Crowley, and Karen Knutson, 81–101. Mahwah, N.J.: Erlbaum, 2002.

Engström, Yngo. *Learning by Expanding: An Activity Theoretical Approach to Developmental Research*. Helsinki, Finland: Orienta-Konsultit, 1987.

Evans, G. "Learning and the Physical Environment." In *Public Institutions for Personal Learning*, edited by John H. Falk and Lynn D. Dierking, 119–26. Washington, D.C.: American Association of Museums, 1995.

Exploratorium. *Electronic Guidebook Forum Report*. San Francisco: Author, January 13–14, 2005.

Facer, Keri. "What Do We Mean by the Digital Divide? Exploring the Roles of Access, Relevance and Resource Networks." In *The Digital Divided*. Coventry, UK: Becta,

February 2002. www.becta.org.uk/page_documents/research/digidivseminar.pdf (accessed May 28, 2007).

Facer, Keri, Richard Joiner, Danaë Stanton, Josephine Reid, Richard Hull, and David S. Kirk. "Savannah: Mobile Gaming and Learning?" *Journal of Computer Assisted Learning* 20, no. 6 (2004): 399–409.

Falk, John H. "Assessing the Impact of Exhibit Arrangement on Visitor Behavior and Learning." *Curator* 36, no. 2 (1993): 1–15.

———. "Field Trips: A Look at Environmental Effects on Learning." *Journal of Biological Education* 17, no. 2 (1983): 134–42.

———. "The Impact of Visit Motivation on Learning: Using Identity as a Construct to Understand the Visitor Experience." *Curator* 49, no. 2 (2006): 151–66.

———. "Museums as Institutions for Personal Learning." *Daedalus* 128, no. 3 (1999): 259–75.

———. "Testing a Museum Exhibition Design Assumption: Effect of Explicit Labeling of Exhibit Clusters on Visitor Concept Development." *Science Education* 81, no. 6 (1997): 679–88.

Falk, John H., and L. Adelman. "Investigating the Impact of Prior Knowledge, Experience and Interest on Aquarium Visitor Learning." *Journal of Research in Science Teaching* 40, no. 2 (2003): 163–76.

Falk, John H., and J. D. Balling. "The Field Trip Milieu: Learning and Behavior as a Function of Contextual Events." *Journal of Educational Research* 76, no. 1 (1982): 22–28.

Falk, John H., P. Brooks, and R. Amin. "Investigating the Long-Term Impact of a Science Center on Its Community: The California Science Center L.A.S.E.R. Project." In *Free-Choice Science Education: How We Learn Science outside of School*, edited by John H. Falk, 115–32. New York: Teachers College Press, 2001.

Falk, John H., and Lynn D. Dierking. *Learning from Museums: Visitor Experiences and the Making of Meaning.* Walnut Creek, Calif.: AltaMira, 2000.

———. *The Museum Experience.* Washington, D.C.: Whalesback Books, 1992.

———, eds. *Public Institutions for Personal Learning.* Washington, D.C.: American Association of Museums, 1995.

Falk, John H., J. E. Heimlich, and K. Bronnenkant. "The Identity-Related Motivations of Adult Zoo and Aquarium Visitors." *Curator*, forthcoming.

Falk, John H., W. W. Martin, and J. D. Balling. "The Novel Field Trip Phenomenon: Adjustment to Novel Settings Interferes with Task Learning." *Journal of Research in Science Teaching* 15 (1978): 127–34.

Falk, John H., Theanno Moussouri, and D. Coulson. "The Effect of Visitors' Agendas on Museum Learning." *Curator* 41, no. 2 (1998): 106–20.

Falk, John H., C. Scott, Lynn D. Dierking, L. Rennie, and M. Cohen-Jones. "Interactives and Visitor Learning." *Curator* 47 (2004): 171–98.

Falk, John H., and M. Storksdieck. "Learning Science from a Leisure Experience: Understanding the Long-Term Meaning Making of Science Center Visitors." *Journal of Research in Science Teaching*, forthcoming.

———. "Using the Contextual Model of Learning to Understand Visitor Learning from a Science Center Exhibition." *Science Education* 89 (2005): 744–78.

Faux, Fern, Angela McFarlane, Neil Rocho, and Keri Facer. *Learning with Handheld Technologies: A Handbook from Futurelab.* Bristol, UK: Futurelab, 2006. www.futurelab.org.uk/research/handbooks/05_01.htm (accessed May 22, 2007).

Filippini-Fantoni, Silvia. "Museums with a Personal Touch." In *EVA 2003 London: Conference Proceedings*, edited by J. Hemsley et al., 25.1–25.10. London, 2003.

Filippini-Fantoni, Silvia, and Jonathan Bowen. "Bookmarking in Museums: Extending the Museum Experience beyond the Visit?" In *Museums and the Web 2007: Proceedings*, edited by Jennifer Trant and David Bearman. Toronto: Archives and Museum Informatics, 2007. www.archimuse.com/mw2007/papers/filippini-fantoni/filippini-fantoni.html (accessed August 13, 2007).

Filippini-Fantoni, Silvia, and Nancy Proctor. "Evaluating the Use of Mobile Phones for an Exhibition Tour at the Tate Modern: Dead End or the Way Forward?" In *EVA London 2007: Conference Proceedings*, edited by Jonathan P. Bowen et al., 8.1–8.11. London, 2007.

Fisher, Susie. "An Evaluation of Learning on the Move and Science Navigator: Using PDAs in Museum, Heritage and Science Centre Settings." Bristol, UK: Nesta Report, 2003.

———. "An Evaluation of Learning on the Move and Science Navigator: Using PDAs in Museum, Heritage and Science Centre Settings." Bristol, UK: Nesta Report, 2005.

———. "Multimedia and BSL Tours at Tate Modern: Stage 2 Qualitative Research." Unpublished evaluation report by the Susie Fisher Group for Tate Modern, London, February 2004.

Fleck, Margaret, Marco Frid, Tim Kindberg, Rakhi Rajani, Eamonn O'Brien-Strain, and Mirjana Spasojevic. "From Informing to Remembering: Deploying a Ubiquitous System in an Interactive Science Museum." *Pervasive Computing* 1, no. 2 (2002): 13–21.

Flickr. Home page. www.flickr.com/.

———. "World Map." www.flickr.com/map.

Fondo Unico. "PEACH: Personal Experience with Active Cultural Heritage." http://peach.itc.it.

Franklin, M. B., R. C. Becklen, and C. L. Doyle. "The Influence of Titles on How Paintings Are Seen." *Leonardo* 26 (1993): 103–8.

Friedman, A. J. "Expanding Audiences: The Audio Tour Access Project at the New York Hall of Science." *Dimensions* (2000): 7–8.

Fritsch, Juliette. "Thinking about Bringing Web Communities into Galleries and How It Might Transform Perceptions of Learning in Museums." Paper presented at the Museum as Social Laboratory: Enhancing the Object to Facilitate Social Engagement and Inclusion in Museums and Galleries seminar, Arts and Humanities Research Council Seminar series, London, March 2007.

Gallo, M. A. "The Effect of Information on the Interpretation of Artwork." Unpublished PhD diss., Rutgers University, New Brunswick, N.J., 2004.

Gammon, Benjamin. "Everything We Currently Know about Making Visitor-Friendly Mechanical Interactive Exhibits." *Informal Learning Review* 39 (1999): 1–13.

———. "Visitors' Use of Computer Exhibits." *Informal Learning Review* 38 (1999): 10–13.

Gardner, Howard. *Frames of Mind: The Theory of Multiple Intelligences.* New York: Basic Books, 1983.

Gelman, Rochel, Christine M. Massey, and Mary McManus. "Characterizing Supporting Environments for Cognitive Development: Lessons from Children in a

Museum." In *Perspectives on Socially Shared Cognition*, edited by Lauren B. Resnick, John M. Levine and Stephanie D. Teasley, 226–56. Washington, D.C.: American Psychological Association, 1991.

Gill, C. J. "Invisible Ubiquity: The Surprising Relevance of Disability Issues in Evaluation." *American Journal of Evaluation* 20, no. 2 (1999): 279–89.

Giusti, Ellen. "Expedition Audio Guides: Treasures from the Museum's Permanent Collections." Unpublished evaluation report for the American Museum of Natural History, 1995.

———. "Summative Evaluation of Audio Guide for Blind and Low Vision Visitors at the New York Hall of Science." Unpublished evaluation report, 2000.

Giusti, Ellen, and J. Watt. "Ping! System Final Evaluation Report." Unpublished summative evaluation report, 2007.

Goldman, Kate Haley. "Cell phones and Exhibitions 2.0: Moving beyond the Pilot Stage." Paper presented at Museums and the Web 2007: Selected Papers from an International Conference, San Francisco, April 2007. www.archimuse.com/mw2007/papers/haleyGoldman/haleyGoldman.html (accessed August 15, 2007).

Goldman, Kate Haley, L. M. Adelman, John H. Falk, K. Owen, K. Buchner, and K. Burtnyck. *Conservation Impact Study National Aquarium in Baltimore: Six-Month Re-contact Study—Technical Report*. Annapolis, Md.: Institute for Learning Innovation, 2001.

Google. "Google Mobile." www.google.co.uk/gmm (accessed June 26, 2008).

Gottlieb, Halina. *Visitor Focus in 21st Century Museums*. Stockholm: Interactive Institute, 2006.

Gottlieb, Halina, Helena Simonsson, S. Lindberg, and L. Asplund. "Audio Guides in Disguise—Introducing Natural Science for Girls." Paper presented at Re-thinking Technology for Museums: Towards a New Understanding of People's Experience in Museums, Limerick, Ireland, June 2005.

Gottlieb, Halina, Helena Simonsson, and H. Öjmyr. "Virtual Touch of a Sculpture." Paper presented at *EVA*, London, 2005.

Graburn, Nelson H. "The Museum and the Visitor Experience." In *The Visitor and the Museum*, 5–32. Seattle: 72nd Annual Conference of the American Association of Museums, 1977.

Greeves, Margaret. "Help at Hand: Working with Handheld Guides." Paper presented at Help at Hand: Working with Handheld Guides Conference, London, June 2006.

Grey, A., T. Gardom, and C. Booth. *Saying It Differently: A Handbook for Museums Refreshing Their Display*. London: London Museums Hub, 2006.

Griffin, Jeanette. "Research on Students and Museums: Looking More Closely at the Students in School Groups." *Science Education* 88, suppl. I (2004): S59–S70.

———. "Learning to Learn in Informal Science Settings." *Research in Science Education* 24, no. 1 (1994): 121–28.

———. "School-Museum Integrated Learning Experiences in Science: A Learning Journey." Unpublished PhD diss., University of Technology, Sydney, 1998.

Gyllenhaal, Eric D., and Deborah L. Perry. "Doing Something about the Weather: Summative Evaluation of Science Museum of Minnesota's *Atmospheric Explorations* Computer Interactives." *Current Trends in Audience Research and Evaluation* 11 (1998). www.selindaresearch.com/GyllenhaalAndPerry1998 Weather.pdf (accessed November 4, 2007).

Halliday, M. "Towards a Language-Based Theory of Learning." *Linguistics and Education* 5 (1993): 93–116.

Halloran, J., E. Hornecker, Geraldine Fitzpatrick, D. Millard, and M. Weal. "The Chawton House Experience: Augmenting the Grounds of a Historic Manor House." Paper presented at Re-thinking Technology for Museums: Towards a New Understanding of People's Experience in Museums, Limerick, Ireland, June 2005.

Ham, S. H. "Cognitive Psychology and Interpretation: Synthesis and Application." In *The Educational Role of the Museum*, edited by Eilean Hooper-Greenhill. 2nd ed. New York: Routledge, 1999.

Hayward, D. G., and M. Brydon-Miller. "Spatial and Conceptual Aspects of Orientation: Visitor Experiences at an Outdoor History Museum." *Journal of Environmental Systems* 13, no. 4 (1984): 317–32.

Heath, Christian, Dirk vom Lehn, and Jonathan Osborne. "Interaction and Interactives: Collaboration and Participation with Computer-Based Exhibits." *Public Understanding of Science* 14, no. 1 (2005): 91–101.

Hedges, A. "Human-Factor Considerations in the Design of Museums to Optimize Their Impact on Learning." In *Public Institutions for Personal Learning*, edited by

John H. Falk and Lynn D. Dierking, 105–18. Washington, D.C.: American Association of Museums, 1995.

Hein, George E. *The Constructivist Museum*, 21–23. Group for Education in Museums, 1995.

———. *Learning in the Museum*. London: Routledge, 1998.

Hickling, Alfred. "Block Beuys." *Guardian*, November 29, 2004.

Higgins, Peter. "The Design Zone." *Museum Practice*, no. 22 (2003): 30–33.

Hinton, Morna. "The Victoria and Albert Museum Silver Galleries II: Learning Style and Interpretation Preference in the Discovery Area." *Museum Management and Curatorship* 17, no. 3 (1998): 253–94.

Holland, D., W. Lachicotte Jr., D. Skinner, and C. Cain. *Identity and Agency in Cultural Worlds*. Cambridge, Mass.: Harvard University Press, 1998.

Hood, M. "Staying Away: Why People Choose Not to Visit Museums." *Museum News* 61, no. 4 (1983): 50–57.

Hooper-Greenhill, Eilean, ed. *The Educational Role of the Museum*. 2nd ed. New York: Routledge, 1999.

———. *Museums and Their Visitors*. London: Routledge, 1994.

———. *Museums and the Shaping of Knowledge*. London: Routledge, 1992.

Hsi, Sherry. "I-Guides in Progress: Two Prototype Applications for Museum Educators and Visitors using Wireless Technologies to Support Informal Science Learning." In *Proceedings of the 2nd IEEE International Workshop on Wireless and Mobile Technologies in Education*, 187–92. JungLi, Taiwan, 2004.

———. "A Study of User Experiences Mediated by Nomadic Web Content in a Museum." *Journal of Computer Assisted Learning* 19, no. 3 (2003): 308–19.

Hyde-Moyer, S. "The PDA Tour: We Did It; So Can You." In *Museums and the Web 2006: Proceedings*, edited by David Bearman and Jennifer Trant. Toronto: Archives and Museum Informatics, 2006. www.archimuse.com/mw2006/papers/hyde-moyer/hyde-moyer.html (accessed June 26, 2008).

ICHIM. "Exploration of Some of Gauguin's Artworks by the Museum Wearable Adapted from the Digital Artistic Reproductions (DAR) Conceived by the Artist Etienne Trouvers." Paris, September 21–21, 2005. www.ichim.org/ichim05/jahia/Jahia/pid/647.html (accessed June 26, 2008).

Intracom. "ARCHEOGUIDE: Augmented Reality-Based Cultural Heritage On-site Guide." http://archeoguide.intranet.gr (accessed June 26, 2008).

Jarvis, Tina, and Anthony Pell. "Factors Influencing Elementary School Children's Attitudes towards Science before, during and after a Visit to the UK National Space Centre." *Journal of Research in Science Teaching* 42, no. 1 (2005): 53–83.

Jewish Museum. "Collection Overview." www.thejewishmuseum.org/Collection Overview (accessed June 26, 2008).

Johnsson, E. *Teachers' Ideas about Learning in Museums.* London: London Museums Hub, 2003.

Judd, David, Morna Hinton, and Frances Lloyd-Baynes. "Interpretation in the Galleries." In *Creating the British Galleries at the V&A: A Study in Museology,* edited by Christopher Wilk and Nick Humphrey, 145–63. London: Victoria and Albert Museum Publications, 2004.

Karp, Ivan, and Steven D. Lavine, eds. *Exhibiting Cultures: The Poetics and Politics of Museum Display.* Washington, D.C.: Smithsonian Institution Press, 1991.

Keen, Andrew. *The Cult of the Amateur: How Today's Internet Is Killing Our Culture and Assaulting Our Economy.* London: Nicholas Brealey Publishing, 2007.

Kennedy, Randy. "With Irreverence and an iPod: Recreating the Museum Tour." *New York Times,* May 28, 2005, A1. www.nytimes.com/2005/05/28/arts/design/28podc .html?ex=1274932800&en=db1ced6873dcc4b6&ei=5090&partner=rssuserland&e mc=rss%20 (accessed August 6, 2007).

Koran, J. J., Jr., M. L. Koran, Lynn D. Dierking, and J. Foster. "Using Modeling to Direct Attention in a Natural History Museum." *Curator* 31, no. 1 (1988): 36–42.

Korn, Randi, et al. *Matthew Barney: Drawing Restraint Interactive Educational Technologies and Interpretation Initiative Evaluation.* San Francisco: SFMOMA, 2006. www.sfmoma.org/whoweare/research_projects/barney/RKA_2006_ SFMOMA_Barney_distribution.pdf (accessed August 10, 2007).

Kubota, C.A., and R. G. Olstad. "Effects of Novelty-Reducing Preparation on Exploratory Behavior and Cognitive Learning in a Science Museum Setting." *Journal of Research in Science Teaching* 28, no. 3 (1991): 225–34.

Laing, A. *In Trust for the Nation.* London: National Trust, 1995.

Lancaster University. "The GUIDE Project." www.guide.lancs.ac.uk.

Laurillard, Diana. *Rethinking University Teaching: A Framework for the Effective Use of Learning Technologies.* 2nd ed. London: Routledge Falmer, 2002.

Laurillau, Y., and F. Paternò. "Supporting Museum Co-visits Using Mobile Devices." In *Proceedings of MobileHCI 04,* 451–55. Berlin: Lecture Notes in Computer Science. http://giove.cnuce.cnr.it/pdawebsite/CiceroPublications.html (accessed November 2005).

Lebeau, R. B., P. Gyamfi, K. Wizevich, and E. H. Koster. "Supporting and Documenting Choice in Free-Choice Science Learning Environments." In *Free-Choice Science Education: How We Learn outside of School,* edited by John H. Falk, 133–48. New York: Teachers College Press, 2001.

Leinhardt, Gaea, Kevin Crowley, and Karen Knutson, eds. *Learning Conversations in Museums.* Mahwah, N.J.: Erlbaum, 2002.

Leinhardt, Gaea, C. Tittle, and Karen Knutson. "Talking to Oneself: Diary Studies of Museum Visits." In *Learning Conversations in Museums,* edited by Gaea Leinhardt, Kevin Crowley, and Karen Knutson. London: Erlbaum, 2002.

Lonsdale, Peter, C. Baber, and Mike Sharples. "A Context Awareness Architecture for Facilitating Mobile Learning." Paper presented at Mlearn 2003—Learning with Mobile Devices, London, May 19–20, 2003.

Lonsdale, Peter, C. Baber, Mike Sharples, W. Byrne, T. N. Arvanitis, and Russell Beale. "Context Awareness for MOBIlearn: Creating an Engaging Learning Experience in an Art Museum." In *Mobile Learning Anytime Everywhere: A Book of Papers from MLEARN 2004,* edited by Jill Attewell and C. Savill-Smith, 115–18. London: Learning and Skills Development Agency, 2004.

Luke, J., M. Cohen Jones, Lynn D. Dierking, M. Adams, and John H. Falk. *The Impact of Museum Programs on Youth Development and Family Learning: The Children's Museum, Indianapolis Family Learning Initiative—Technical Report.* Annapolis, Md.: Institute for Learning Innovation, 2002.

Luke, J., and J. Stein. "Walker Art Center: Summative Evaluation of Interpretive Experiences Newly Installed within the Permanent Collection." Unpublished technical report. Annapolis, Md.: Institute for Learning Innovation, 2006.

Lyons, Leilah, Joseph Lee, Christopher Quintana, and Elliot Soloway. "MUSHI: A Multi-device Framework for Collaborative Inquiry Learning." Paper presented at

Proceedings of the 7th International Conference on Learning Sciences. Bloomington, Ind., 2006, 453–59.

MacDonald, Sharon. *Behind the Scenes at the Science Museum.* New York: Berg, 2002.

Manning A., and G. Sims. "The Blanton iTour—An Interactive Handheld Museum Guide Experiment." In *Museums and the Web 2004: Proceedings,* edited David Bearman and Jennifer Trant. Toronto: Archives and Museum Informatics, 2004. www.archimuse.com/mw2004/papers/manning/manning.html (accessed June 26, 2008).

Marti, P. *The User Evaluation.* Restricted report. Siena, Italy: Università degli Studi di Siena, 2000.

Marti, P., et al. "HIPS: Hyper-Interaction within Physical Space." In *Proceedings of the IEEE International Conference on Multimedia Computing and Systems.* Vol. 2. Washington, D.C.: IEEE Computer Society, 1999.

Martin, L. M. "An Emerging Research Framework for Studying Informal Learning and Schools." *Science Education* 88 (2004): S71–S82.

Mayer, Richard E. *Multimedia Learning.* Cambridge, UK: Cambridge University Press, 2001.

McClamrock, R. *Existential Cognition.* Chicago: University of Chicago Press, 1995.

McIntyre, Morris Hargreaves. "CIPs in Context: Understanding Visitor Usage of Computer Information Points at the Science Museum." Unpublished evaluation report for the Science Museum, London, 2006.

———. *Engaging or Distracting: Visitor Responses to the Interactives in the V&A British Galleries.* Evaluation report for the Victoria and Albert Museum, London, 2003. www.vam.ac.uk/files/file_upload/5877_file.pdf (accessed May 21, 2007).

McManus, Paulette. "It's the Company You Keep: The Social Determination of Learning-Related Behavior in a Science Museum." *International Journal of Museum Management and Curatorship* 53 (1987): 43–50.

———. "Topics in Museums and Science Education." *Studies in Science Education* 20 (1992): 157–82.

Medved, M. I. "Remembering Exhibits at Museums of Art, Science and Sport." Unpublished PhD diss., University of Toronto, 1998.

Meisner, Robin, Dirk vom Lehn, Christian Heath, Alexandra Burch, Ben Gammon, and Molly Reisman. "Exhibiting Performance: Co-participation in Science Centres and Museums." *International Journal of Science Education* 29, no. 12 (2007): 1531–35.

Merriman, N. *Beyond the Glass Case.* Leicester, UK: Leicester University Press, 1991.

Meszaros, Cheryl. "Now THAT Is Evidence: Tracking Down the Evil 'Whatever' Interpretation." *Visitor Studies Today* 9, no. 3 (Winter 2006): 10–15.

Metfinder. "Metfinder: A Handheld Solution for Independent Exploration and Discovery in the Museum." Paper presented at MCN Conference, Pasadena, Calif., November 2006. http://72.5.117.137/conference/mcn2006/SessionPapers/Metfinder.pdf (accessed June 26, 2008).

Miles, R. S. "Museum Audiences." *International Journal of Museum Management and Curatorship* 5 (1986): 73–80.

Miller, George A. "The Magical Number Seven, Plus or Minus Two: Some Limits on Our Capacity for Processing Information." *Psychological Review* 63 (1956): 81–97. www.well.com/user/smalin/miller.html (accessed November 5, 2007).

Millis, K. "Making Meaning Brings Pleasure: The Influence of Titles on Aesthetic Experiences." *Emotion* 1 (2001): 320–29.

Moussouri, Theanno. "Family Agendas and Family Learning in Hands-On Museums." Unpublished PhD diss., University of Leicester, UK, 1997.

Mulholland, Paul, T. Collins, and Z. Zdrahal. "Bletchley Park Text: Using Mobile and Semantic Web Technologies to Support the Post-visit Use of Online Museum Resources." *Journal of Interactive Media in Education* (December 2005). www-jime.open.ac.uk/2005/21/ (accessed November 5, 2007).

Naismith, Laura, Peter Lonsdale, Giasemi Vavoula, and Mike Sharples. *Literature Review in Mobile Technologies and Learning.* Bristol, UK: Futurelab, 2005. www.futurelab.org.uk/research/reviews/reviews_11_and12/11_01.htm (accessed May 21, 2007).

National Museums Directors' Conference. *A Netful of Jewels: New Museums in the Learning Age.* London: Author, 1999.

Norman, Donald A. *The Design of Everyday Things.* New York: Basic Books, 2002.

Novey, L. T., and T. E. Hall. "The Effect of Audio Tours on Learning and Social Interaction: An Evaluation at Carlsbad Caverns National Park." *Science Education* (2006), 260–77.

Ogbu, John. "The Influence of Culture on Learning and Behavior." In *Public Institutions for Personal Learning: Establishing a Research Agenda*, edited by John H. Falk and Lynn D. Dierking, 79–96. Washington, D.C.: American Association of Museums, 1995.

Ogden, J. L., D. G. Lindburg, and T. L. Maple. "The Effects of Ecologically Relevant Sounds on Zoo Visitors." *Curator* 36, no. 2 (1993): 147–56.

Öjmyr, H. "Vad minns barn efter ett konstmuseibesök?" Unpublished report. Stockholm: Interactive Institute, 2000.

O'Neill, M. "Interpretation Philosophy behind Kelvingrove." Glasgow Museums, 2006. www.glasgow.gov.uk/en/Visitors/MuseumsGalleries/KelvingroveMedia Office/NewKelvingrovePhilosophy/Kelvingrovephilosophyinterpretation.htm (accessed November 1, 2007).

Packer, J. "Learning for Fun: The Unique Contribution of Educational Leisure Experiences." *Curator* 49, no. 3 (2006): 329–44.

Packer, J., and R. Ballantyne. "Motivational Factors and the Visitor Experience: A Comparison of Three Sites." *Curator* 45 (2002): 183–98.

Paris, S., ed. *Perspectives on Object-Centered Learning in Museums*. Mahwah, N.J.: Erlbaum, 2002.

Parry, Ross. *Recoding the Museum: Digital Heritage and the Technologies of Change.* London: Routledge, 2007.

Pask, Gordon. "Minds and Media in Education and Entertainment: Some Theoretical Comments Illustrated by the Design and Operation of a System for Exteriorizing and Manipulating Individual Theses." In *Progress in Cybernetics and Systems Research*, edited by R. Trappl and Gordon Pask, 38–50. Vol. 4. Washington, D.C.: Hemisphere Publishing Corporation, 1975.

Paxman, Jeremy. "What's Wrong with TV?" James MacTaggart Memorial Lecture, MediaGuardian Edinburgh International Television Festival, August 2007. www.telegraph.co.uk/news/main.jhtml?xml=/news/2007/08/25/npaxspeech125.xm l&page=3 (accessed November 1, 2007).

Pekarik, Andrew J. "To Explain or Not to Explain." *Curator* 47 (2004): 12–18.

Pekarik, Andrew J., Zahava D. Doering, and Adam Bickford. "Visitors' Role in an Exhibition Debate: Science in American Life." *Curator* 42, no. 2 (1999): 117–29.

Pekarik, Andrew J., Zahava D. Doering, and David A. Karns. "Exploring Satisfying Experiences in Museums." *Curator* 42, no. 2 (1999): 152–73.

Petrelli, D., E. Not, M. Sarini, C. Strapparava, O. Stock, and M. Zancanaro. "HyperAudio: Location-Awareness + Adaptivity." In *ACM SIGCHI '99 Extended Abstracts*, 21–22. Pittsburgh, May 1999. http://citeseer.ist.psu.edu/petrelli99 hyperaudio.html (accessed June 26, 2008).

Pierroux, Palmyre. "The Language of Contextualism and Essentialism in Museum Education." Paper presented at Re-thinking Technology for Museums: Towards a New Understanding of People's Experience in Museums, Limerick, Ireland, June 2005.

Pitman, Bonnie. "Serving Visitors with Choices: How Far Can Art Museums Stretch?" Presentation on a panel at American Association of Museums Annual Meeting, Chicago, May 2007.

Prentice, R., A. Davies, and A. Beeho. "Seeking Generic Motivations for Visiting and Not Visiting Museums and Like Cultural Attractions." *Museum Management and Curatorship* 6 (1997): 45–70.

Proctor, Nancy. "Access in Hand: Providing Deaf and Hard-of-Hearing Visitors with On-Demand, Independent Access to Museum Information and Interpretation through Handheld Computers." Paper presented at ICHIM, Berlin, 2004. www.ichim.org/ichim04/contenu/PDF/4324_Proctor.pdf.

———. *Antenna Audio at the Tate Modern*. London: Antenna Audio, 2004.

———. "Off Base or on Target? Pros and Cons of Wireless and Location-Aware Applications in the Museums." Paper presented at ICHIM, Paris, 2005.

Proctor, Nancy, and Jane Burton. "Tate Modern Multimedia Tour Pilots 2002–2003." Unpublished evaluation report from Tate Modern, London, 2003.

Proctor, Nancy, Jane Burton, and Chris Tellis. "The State of the Art in Museum Handhelds in 2003." In *Museums and the Web 2003: Selected Papers from an International Conference*, Charlotte, N.C., March 2003. www.archimuse.com/mw2003/papers/proctor.html (accessed June 26, 2008).

Puig, Vincent, and Xavier Sirven. "*Lignes De Temps*: Involving Cinema Exhibition Visitors in Mobile and On-line Film Annotation." Paper presented at Museums and the Web 2007, San Francisco, April 2007. www.archimuse.com/mw2007/papers/puig/puig.html (accessed November 5, 2007).

Ravelli, L. *Museum Texts: Communication Frameworks*. London: Routledge, 2006.

Ravenscroft, A. "Designing Argumentation for Conceptual Development." *Computers and Education* 34 (2000): 241–55.

Reich, C., and A. Lindgren-Streicher. "Universal Design Literature Review." Unpublished report for the Museum of Science, Boston, 2005.

Rennie, L., and T. P. McClafferty. "Science Centres and Science Learning." *Studies in Science Education* 22 (1996): 53–98.

Resource. *Information and Communications Technology and the Development of Museums, Archives and Libraries: A Strategic Plan for Action*. Draft for consultation. London: Author, 2001.

———. *Renaissance in the Regions: A New Vision for England's Museums*. London: Regional Museums Task Force for Resource, the Council for Museums, Libraries and Archives, 2001.

Reynolds, Ellen, with input from Lynda Kelly. "Handheld Report." Unpublished report for AMARC, 2005.

Rheingold, Howard. *Smart Mobs: The Next Social Revolution*. Cambridge, Mass.: Basic Books, 2002.

Rogoff, Irit. *Museum Culture: Histories, Discourses, Spectacles*. London: Routledge, 1994.

Roschelle, Jeremy. "Learning in Interactive Environments: Prior Knowledge and New Experience." In *Public Institutions for Personal Learning*, edited by John H. Falk and Lynn D. Dierking, 37–51. Washington, D.C.: American Association of Museums, 1995.

Rosenfeld, S. "Informal Education in Zoos: Naturalistic Studies of Family Groups." Unpublished PhD diss., University of California, Berkeley, 1980.

Ross, S., M. Donnelly, and M. Dobreva. *Emerging Technologies for the Cultural and Scientific Heritage Sector*. DigiCULT Technology Watch Report 2. Salzburg, Austria: European Commission, 2004.

Russell, P. A., and S. Milne. "Meaningfulness and Hedonic Value of Paintings: Effects of Title." *Empirical Studies of the Arts* 15 (1997): 61–73.

Säljö, R., *Lärande i praktiken. Ett sociokulturellt perspektiv*. Stockholm: Prisma, 2000.

Samis, Peter. "Gaining Traction in the Vaseline: Visitor Response to a Multi-track Interpretation Design for Matthew Barney: Drawing Restraint." Paper presented at

Museums and the Web, San Francisco, April 2007. www.archimuse.com/mw2007/
papers/samis/samis.html (accessed June 26, 2008).

Samis, Peter, and S. Pau. "'Artcasting' at SFMOMA: First-Year Lessons, Future
Challenges for Museum Podcasters." Paper presented at Museums and the Web,
Albuquerque, N.Mex., April 2007. www.archimuse.com/mw2006/papers/samis/
samis.html (accessed August 6, 2007).

Sandifer, Cody. "Technological Novelty and Open-Endedness: Two Characteristics of
Interactive Exhibits that Contribute to the Holding of Visitor Attention in a
Science Museum." *Journal of Research in Science Teaching* 40, no. 2 (2003): 121–37.

Schaubel, L., D. Banks, G. D. Coates, L. M. W. Martin, and P. Sterling. "Outside the
Classroom Walls: Learning in Informal Environments." In *Innovations in Learning*,
edited by L. Schauble and R. Glaser, 5–24. Mahwah, N.J.: Erlbaum, 1996.

Schön. D. *Educating the Reflective Practitioner: Toward a New Design for Teaching and
Learning in the Professions.* San Francisco: Jossey-Bass, 1987.

Schwarzer, Marjorie. "Art and Gadgetry: The Future of the Museum Visit." *Museum
News* 68 (July/August 2001): 36–41.

Science Museum, London. "Energy—Fuelling the Future." www.sciencemuseum
.org.uk/on-line/energy/site/about.asp (accessed May 27, 2007).

———. "Energy: Fuelling the Future, Summative Evaluation." Unpublished
evaluation report, 2005.

———. "Energy Hall." www.sciencemuseum.org.uk/visitmuseum/galleries/
energy_hall.aspx (accessed August 9, 2007).

———. "Nanotechnology—Small Science, Big Deal." www.sciencemuseum.org.uk/
antenna/nano/ (accessed August 9, 2007).

———. "Nanotechnology—Small Science, Big Deal, Summative Evaluation."
Unpublished evaluation report, 2005.

———. "Soundbytes Audio-Tour, Summative Evaluation." Unpublished evaluation
report, 2003.

Scottish Museums Council. *Museums, Galleries and Digitisation: Current Best Practice
and Recommendations on Measuring Impact.* Edinburgh: Author, 2005.

Screven, Chandler. *The Measurement and Facilitation of Learning in the Museum Environment: An Experimental Analysis.* Washington, D.C.: Smithsonian Institute Press, 1974.

Sellen, Abigail J., and Richard H. Harper. *The Myth of the Paperless Office.* Cambridge, Mass.: MIT Press, 2002.

Serrell, Beverly. *Exhibit Labels: An Interpretive Approach.* Walnut Creek, Calif.: AltaMira, 1996.

———. "Paying Attention: The Duration and Allocation of Visitors' Time in Museum Exhibitions." *Curator* 40, no. 2 (1997): 108–25.

———. *Paying Attention: Visitors and Museum Exhibitions.* Washington, D.C.: American Association of Museums, 1998.

Serrell, Beverly, and Britt Raphling. "Computers on the Exhibit Floor." *Curator* 35, no. 3 (1992): 181–89.

Serrell, Beverly, and S. Tokar. "New York Hall of Science Random-Access Audio Tour Project." Unpublished evaluation report, 1997.

Sharples, Mike. "Learning as Conversation: Transforming Education in the Mobile Age." Paper presented at the Conference on Seeing, Understanding, Learning in the Mobile Age, Budapest, Hungary, April 2005.

Sharples, Mike, Josie Taylor, and Giasemi Vavoula. "Towards a Theory of Mobile Learning." Paper presented at the 4th World Conference on Mlearning, Cape Town, South Africa, 2005. www.mlearn.org.za/CD/papers/Sharples-%20Theory%20of%20Mobile.pdf (accessed June 30, 2007).

Siemens, George. "Connectivism: A Learning Theory for the Digital Age." December 4, 2004. www.elearnspace.org/Articles/connectivism.htm (accessed June 26, 2008).

———. "Connectivism: Museums as Learning Ecologies." Presentation to the Canadian Heritage Information Network's Roundtable on e-Learning, March 2006. www.elearnspace.org/media/CHIN/player.html (accessed August 30, 2007).

Silverman, L. "Making Meaning Together." *Journal of Museum Education* 18, no. 3 (1993): 7–11.

Smith, Jeffrey K. "Analysis of 'Key to the Met' Audio Program." Unpublished report. New York: Metropolitan Museum of Art, 2000.

————. "Report of Users and Non-users of Audio Tours at the Jewish Museum." Unpublished report. New York: Jewish Museum, 2001.

Smith, Jeffrey K., and D. W. Carr. "In Byzantium." *Curator* 44, no. 4 (2001): 335–54.

Smith, Jeffrey K., and L. F. Smith. "Spending Time on Art." *Empirical Studies in the Arts* 19, no. 2 (2001): 229–36.

Smith, Jeffrey K., I. Waszkielewicz, K. Potts, and B. K. Smith. "Visitors and the Audio Program: An Investigation into the Impact of the Audio Guide Program at the Whitney Museum of American Art." Internal report. New York: Whitney Museum of American Art, 2004.

Smith, M. "Paulo Friere." In *Encyclopedia of Informal Education.* 2005. www.infed.org/thinkers/et-freir.htm (accessed August 29, 2007).

Spalding, S. *The Poetic Museum: Reviving Historic Collections.* Munich, Germany: Prestel, 2002.

Spasojevic, Mirjana, and Tim Kindberg. "Evaluating the CoolTown User Experience." Paper presented at the Workshop on Evaluation Methodologies for Ubiquitous Computing, held at Ubicomp'01, Atlanta, 2001. www.exploratorium.edu/guidebook/eguide_exec_summary_02.pdf (accessed November 6, 2007).

Stainton, C. "Voice and Images: Making Connections Between Identity and Art." In *Learning Conversations in Museums*, edited by Gaea Leinhardt, Kevin Crowley, and Karen Knutson, 213–57. Mahwah, N.J.: Erlbaum, 2002.

Sternberg, R. J., and R. K. Wagner, eds. *Practical Intelligence: Nature and Origins of Competence in the Everyday World.* Cambridge, UK: Cambridge University Press, 1996.

Stevens, Reed, and Roger Hall. "Seeing Tornado: How Video Traces Mediate Visitor Understandings of (Natural?) Phenomena in a Science Museum." *Science Education* 81, no. 6 (1997): 735–47.

Stolterman, Eric. *Utskrifter av 20 intervjuer med systemutvecklare* (UMADP-WPIPCS 43.91). Umeå: Inst. för Informationsbehandling/ADB, Umeå Universite, 1991.

Sweller, John, J. J. G. van Merriënboere, and F. G. W. C. Paas. "Cognitive Architecture and Instructional Design." *Education Psychology Review* 10 (1998): 251–96.

Tallon, Loïc. "Audio Tours in Historic Houses." *Historic Houses Association Journal* (Autumn 2006): 40–41.

———. "Education and the Audio Tour." *Engage* 18 (Winter 2006): 65–69.

———. "The Evolution of the Audio Tour Narrator." *Museums Journal* 106, no. 4. (April 2006): 26–29.

Tate Modern. *Tate Modern Multimedia Tour*, 2004. www.tate.org.uk/modern/multimediatour/reseval.htm (accessed April 11, 2007).

Tellis, Chris. "Multimedia Handhelds: One Device, Many Audiences." In *Museums and the Web 2004: Proceedings*, edited by David Bearman and Jennifer Trant. Toronto: Archives and Museum Informatics, 2004. www.archimuse.com/mw2004/papers/tellis/tellis.html (accessed June 26, 2008).

Temme, J. E. V. "Amount and Kind of Information in Museums: Its Effects on Visitors' Satisfaction and Appreciation of Art." *Visual Arts Research* 18 (1992): 74–81.

Tinio, Pablo L., K. Potts, and Jeffrey K. Smith. "The Museum Tour: Visitors, Guide, and Interactivity." Unpublished manuscript, 2006.

Topalian, R. "Cultural Visit Memory: The Visite+ System Personalization and Cultural Visit Tracking Site." In *Museums and the Web 2005: Proceedings*, edited by David Bearman and Jennifer Trant. Toronto: Archives and Museum Informatics, 2005. www.archimuse.com/mw2005/papers/topalian/topalian.html.

TWResearch. "Evaluation of a Multimedia Guide Accompanying the Frida Kahlo Exhibition." Unpublished evaluation report for Tate Modern, London, 2005.

University of Southampton. "Creating the 'Outdoor Classroom.'" July 14, 2005, www.soton.ac.uk/mediacentre/news/2005/jul/05_137.shtml (accessed March 26, 2008).

Vahey, P., Jeremy Roschelle, and D. Tatar. "Using Handhelds to Link Private Cognition and Public Interaction." *Educational Technology* (May–June 2007): 13–16.

Van Loon, Heleen, Kris Gabriël, Kris Luyten, Daniel Teunkens, Karel Robert, Karin Coninx, and Elke Manshoven. "Supporting Social Interaction: A Collaborative Trading Game on a PDA." In *Museums and the Web 2007: Proceedings*, edited by Jennifer Trant and David Bearman. Toronto: Archives and Museum Informatics, 2007. www.archimuse.com/mw2007/papers/vanLoon/vanLoon.html (accessed August 13, 2007).

Van Moer, E. "Talking about Contemporary Art: The Formation of Preconceptions during a Museum Visit." *International Journal of the Arts in Society* 1, no. 3 (2006): 1–8.

Vavoula, Giasemi. "A Study of Mobile Learning Practices." Deliverable 4.4 for the MOBIlearn project (EU, IST-2001-37440), 2005. www.mobilearn.org/download/ results/public_deliverables/MOBIlearn_D4.4_Final.pdf (accessed August 29, 2007).

Victoria and Albert Museum. "British Galleries." www.vam.ac.uk/collections/ british_galls/index.html (accessed May 27, 2007).

———. "History of the V&A: Your First Visit to the V&A." www.vam.ac.uk/ collections/periods_styles/history/first_visit/index.php (accessed March 27, 2008).

vom Lehn, Dirk, and Christian Heath. "Displacing the Object: Mobile Technologies and Interpretative Resources." Paper presented at ICHIM 03, Paris, September 2003. www.ichim.org/ichim03/PDF/088C.pdf (accessed May 22, 2007).

Vygotsky, L. S. "The Genetic Roots of Thought and Speech." In *Thought and Language*, translated and edited by A. Kozulin, 68–95. Cambridge, Mass.: MIT Press, 1986.

———. *Mind in Society: The Development of Higher Psychological Processes.* Cambridge, Mass.: Harvard University Press, 1978. (Published originally in Russian in 1930.) www.marxists.org/archive/vygotsky/works/mind (accessed November 5, 2007).

Waldrop, J., and M. Stern. *Disability Status: 2000.* 2003. www.census.gov/prod/ 2003pubs/c2kbr-17.pdf (accessed April 11, 2007).

Walker, Kevin. "Visitor-Constructed Personalized Learning Trails." In *Museums and the Web 2007: Proceedings*, edited by Jennifer Trant and David Bearman. Toronto: Archives and Museum Informatics, 2007. www.archimuse.com/mw2007/papers/ walker/walker.html (accessed June 26, 2008).

Waycott, Jenny, Ann Jones, and Eileen Scanlon. "PDAs as Lifelong Learning Tools: An Activity Theory Based Analysis." *Learning, Media and Technology* 30, no. 2 (July 2005): 107–30.

Weinberger, David. *Everything Is Miscellaneous: The Power of the New Digital Disorder.* New York: Times Books/Henry Holt, 2007. www.everythingis miscellaneous.com/ (accessed August 28, 2007).

Weiser, Mark. "The Computer for the 21st Century." *Scientific American* 265, no. 3 (1991): 73–76.

Wells, M., and R. Loomis. "A Taxonomy of Museum Opportunities: Adapting a Model from Natural Resource Management." *Curator* 41 (1998): 254–64.

Wenger, Etienne. "Communities of Practice: Learning as a Social System." *Systems Thinker* (June 1998). www.co-i-l.com/coil/knowledge-garden/cop/lss.shtml (accessed November 5, 2007).

Wertsch, James V. *Voices of the Mind: A Sociocultural Approach to Mediated Action.* Cambridge, UK: Cambridge University Press, 1991.

———. *Vygotsky and the Social Formation of the Mind.* Cambridge, Mass.: Harvard University Press, 1985.

Wetterlund, K., and S. Sayre. "2003 Art Museum Education Programs Survey." www.museum-ed.org/ (accessed May 27, 2007).

Whitney Museum of American Art. "History of the Whitney." www.whitney.org/.

Wiese, Erich. "Experiences with Short Wave Radio Tours in the Hesse Museum at Darmstadt, in Museumskunde." Unpublished letter by Der Deutsche Museumbund, Verlag Walter de Gruyter & Co., Berlin, 1960/1961.

Wikipedia. Home page. http://wikipedia.org.

Williamson-Shaffer, David, and Mitchel Resnick. "'Thick' Authenticity: New Media and Authentic Learning." *Journal of Interactive Learning Research* 10, no. 2 (1999): 195–215.

Wilson, Gillian. "Multimedia Tour Programme at Tate Modern." Paper presented at the Museums and the Web Conference, Washington, D.C., April 2004. www.archimuse.com/mw2004/papers/wilson/wilson.html (accessed May 27, 2007).

Wolins, I., N. Jensen, and R. Ulzheimer. "Children's Memories of Museum Field Trips: A Qualitative Study." *Journal of Museum Education* 17, no. 2 (1992): 17–27.

Woodruff, Allison, Paul M. Aoki, Rebecca E. Grinter, Amy Hurst, Margaret H. Szymanski, and James D. Thornton. "Eavesdropping on Electronic Guidebooks: Observing Learning Resources in Shared Listening Environments." In *Museums and the Web 2002: Selected Papers from an International Conference*, edited by David Bearman and Jennifer Trant, 21–30. Pittsburgh: Archive and Museum Informatics, 2002.

Woodruff, Allison, Paul M. Aoki, Amy Hurst, and Margaret H. Szymanski. "Electronic Guidebooks and Visitor Attention." In *Proceedings of 6th International Cultural Heritage Informatics Meeting*, edited by David Bearman and Jennifer Trant, 437–54. Philadelphia: Archive and Museum Informatics, 2001.

YouTube. "Broadcast Yourself." http://uk.youtube.com/.

Zeeuw, Gerard de. *Coordinated Cooperation and Increasing Competence*. Amsterdam: UVA/OOC, 1990.

Index

access, 97–108, 183; intellectual, 98–99; physical, 100–107; to technology, 190. *See also* visitors
Acoustiguide, xix
American Association of Museums, 23
American Museum of Natural History, xxiv
American Sign Language. *See* language, sign
American Visionary Art Museum, 10
Americans with Disabilities Act, 101. *See also* access; visitors
Antenna Audio, xix, 7, 12
aquaria, xxii, 28, 35
ARCHEOGUIDE, 85
architecture, 22, 24, 79, 128
Art Mobs, 6
audio guides. *See* audio tours
audio recording, 115–66, 119, 158, 162–63, 182
audio tours, 63–76, 167–72, 190; cassette players in, xiii, xvii, xix, 98–99; history of, x, xiii, xix, xxii–xxv, 5–6, 98–99; linear, 5, 67, 99; operation of, 44; psychologists on, 72; random access, xiv, 6, 99, 185; record players in, xix
Avesta Ironwork, 167, 172–73

Baldessari, John, 10
Barber, Thomas, 180
barcode scanning, 86
Barney, Matthew, 12
beacons, 102, 104, 130
Bellows, George, 72, 75
Bletchley Park, 114, 120
Bluetooth, 85–86
bookmarking, 37, 43–44, 83–84, 114–15, 132, 135–39. *See also* extending the visit; RFID tagging
budgets. *See* costs
Burden, Chris, 72–73, 75
Bush, Vannevar, 111
By-Word, xix

cameras: at exhibits, 133, 135,
    137–38; digital, xviii, 152, 163,
    182, 190; digital video, 36; phones
    with, 147, 157–63
cancer, 142
cell phone, 10, 37, 79, 102–106, 112,
    114, 132, 157–61, 182, 187, 190
Centre Pompidou, 11, 115
Chawton House, 83
chiaroscuro, 99
choice and control, 21, 50, 52, 99,
    101, 106, 184
City of Science and Industry, 84
classroom, 147, 156, 163. *See also*
    schools; teachers
cognitive, 50, 89, 137, 156, 161
collaboration, 148, 152. *See also*
    dialogue; social interaciton
Concord Consortium, 130
connectivism, 11, 13
constructivism, 4, 21, 110
content development, 86
context: crossing, 147–65; personal,
    21; physical, 22; sociocultural, 21
Context-Aware Gallery Exploration
    (CAGE), 153–56, 161, 163
Contextual Model of Learning,
    20–24, *23*
control. *See* choice and control
conversation. *See* dialogue
copyright, 86
costs, xx, 86, 90, 161, 189
Creative Time, xxiv
Crissy Field, 127–28
cultural heritage, 80, 175–76, 180,
    187, 189
curator, ix, x, xx, 4, 5, 7, 66, 81, 98,
    111, 117, 157, 164, 182, 186;

authority, 110. *See also* voice;
    museum

Dallas Museum of Art, 12
database, 45, 112, 120, 127
del.icio.us, 164
Denver Art Museum, 10
design, 134, 143–44; interaction, 167
dialogue: and learning, 148–49; and
    technology, 159–60;
digital, xx; inhibition of, 69;
    productive, 142; visitors', xxii, 5,
    109, 117–18, 133, 169; with
    physical environment, 22
D-Day Museum, 157, 159
disabled. *See* access
distraction, 39, 51, 66, 89, 112–13, 133
docent, 126, 133; tours, 65, 66;
    technology for, 139–41
download, 6, 12, 25, 80, 91, 157
Duchamp, Marcel, 6
Dulwich Picture Gallery, 82, 112

earpiece, 48. *See also* headphones;
    speaker
ecology, 128, 170–72
Eliasson, Olafur, 3, 9
elitism, 98
e-mail, 120, 129, 164
Empire State Building, xiv
Energy and Matter Worlds, 137
"Energy—Fuelling the Future"
    exhibit, 36
Energy Hall (Science Museum,
    London), 39
environment. *See* ecology; context
Equator project, 150
Erice, Victor, 11

Experience Music Project, xvii, 80
explainer. *See* docent
Exploratorium, xi, 35, 83, 117,
 125–44
exSpot, 135–39
extending the visit, 26–27, 83–84,
 118, 132–33, 135, 137, 144,
 147–65, 167, 183

families, 48, 50, 118, 144
fatigue, 129, 133. *See also*
 information overload; portability
feedback, 46, 118
Fitzwilliam Museum, 80, 114
flashlight, 172–73
Flickr, 164
Forbidden City, xiv
Forrester Research, 11
free-choice learning, 19, 21–22, 99,
 101, 156
Fritsch, Juliette, 110

gallery, x, xiii, xiv, xx, 6, 48, 64–66,
 76, 79, 83, 111, 119, 148, 153–56;
 conversation, 10. *See also*
 museums; art
Gallo-Roman Museum, 83
Game of Life, 142
games, 36, 37, 45, 81, 83, 141–42,
 150, 171
Gardner, Howard, 4
George Square, 150–53, 161, 163
Getty Museum, 80; GettyGuide, 86
"Ghost Landscape," 127–29, 132
Glenbow Museum, 80
Global Positioning System. *See* GPS
Golden Gate National Recreation
 Area, 127

Google Maps, 149
Gorky, Arshile, 72
GPS, 85, 151–52, 163
Great Blacks in Wax, 82
Grynsztejn, Madeleine, 5
GUIDE, 85

hardware, 90. *See also* cell phone;
 headphones; iPod; MP3 player;
 Personal Digital Assistant
headphones, 48, 133, 154, 170. *See
 also* earpiece; headset; speaker
headset, 151
help, 45, 102
Hewlett-Packard Jornada, 130
HIPS, 85
historical house, 40, 46
HP Labs Palo Alto, 130
Hyperaudio, 85

icons, 88, 151–52
Impressionism, 8
Industrial Revolution, 39
information overload, 10, 50, 87. *See
 also* choice and control; structure
infrared, 130
infrastructure, 80, 85, 143–44, 164,
 183. *See also* network; wireless
inquiry, 4, 9, 35, 127, 129
installations, 87, 128, 139, 172,
 175–76, 190
Institute for Museum and Library
 Services, 8
Intel Labs Seattle, 137
Internet. *See* Web
Internet Archive, 128
interpretation, x, xiv, 5, 9, 13, 37, 51,
 64, 81–82, 84, 91–92, 104, 116,

117, 143, 184–85, 188; abstract art
of, 64; audio, 98, 102;
communities of, 110; digital, 35,
39, 41–42, 48, 51, 112, 114, 132;
mobile, 80, 82, 175; personalized,
37, 99, 115, 158, 187; shared, 148.
See also curator
iPod, x, 79, 86

Jewish Museum, 67, 69–71, 76
Johnson, Samuel, 180

Kahle, Brewster, 128
Kahlo, Frida, 41
Kedleston Hall, 180
Kew Gardens, 112, 119, 120
keypad, 80, 86, 115
Kiarostami, Abbas, 11
kiosk, 16n, 35, 39, 41, 48, 50, 52n1,
79, 164
Kusama, Yayoi, 72, 75

labels: digital, 36, 39–40, 138;
extended object, 5; video, 35; text,
xiii, xviii, xx, xxi, 5, 10, 22, 38, 43,
49, 65–66, 68, 71–72, 74–75,
100–102, 107, 115, 180, 188, 190.
See also text
language: foreign, xvii; sign, 82, 106;
user, xi–xii. *See also* visitors, deaf
Larson, Brad, 16n22
learning styles, xiii, xvii, 4, 6, 35, 47,
81, 182
learning trails. *See* trails
Lignes de Temps, 11
location tracking, 46, 85–86, 88,
150–56, 162
long tail, xvii

longitudinal research, 27
Louvre, xiv
Lund principles, 188

magnetic tape, xiii
maintenance, xix, 80, 86–88, 90
maps, 87–88, 150–53, 162, 190;
audio, 102; digital, 85, 116, 128.
See also Google Maps
McLuhan, Marshall, x
Memex, 111
memory, 44, 111, 120, 128, 132, 157
mental model, 42, 43–44, 52, 109, 120
Mentuhotep II, xxiv
Metropolitan Museum of Art, xxiv,
8, 67–69, 75, 98; MetFinder, 88.
See also de Montebello, Philippe
Milwaukee Public Museum, xxiv
mobile phone. *See* cell phone
MOBIlearn project, 153
de Montebello, Philippe, 98
MP3 player, xviii, xxi, xxiv, 6, 25, 86,
99, 163, 190
multimedia, 79–92, 129–41, 167,
171–73, 175
multiple intelligences theory, 4
Museo Civico, 85
Museum of Modern Art (MoMA), 6,
7
Museum of Science (Boston), 82, 83,
106–107
Museum Wearable, 85
museums: art, 28, 38, 98, 168–72;
science, 28, 36, 51, 87, 99, 102, 147
music, 63, 68–69, 81–82, 163, 171
music player. *See* MP3 player
MyArtSpace, 11, 83, 116, 119,
156–61, 158

"Nanotechnology—Small Science, Big Deal" exhibition, 40
narrative, 35, 109, 111, 117, 129, 168–75, 180, 182, 184–85
National Gallery (London), 39
National Gallery of Art (Washington, D.C.), *xxiii*, xxiv
Nationalmuseum (Stockholm), 167–70
Natural History Museum (London), 80
navigation: in museum, 51, 85–86, 88, 101–106; 150–53, 179; on device, 45, 87. *See also* GPS; maps
network, 104, 141. *See also* infrastructure; wireless
New York Hall of Science, 99–106
New York Times, 6

objectivity, 3–4
operating system, 87
operation: of exhibits, 51; of handheld devices, 42, 45–46
orientation. *See* navigation
outdoors, 127–29
ownership, 132

pacing, 45
Palace of Fine Arts, 128, 132
palmtop. *See* Personal Digital Assistant
participatory simulations. *See* simulations
partnerships, 143
PEACH, 85
performance, 185
Personal Digital Assistant (PDA), xviii, 37, 44, 46, 50, 51, 79, 81, 83, 87, 112, 114, 125, 142, 144, 153–56, 163, 167, 182, 187
personalization, 84–85, 89, 139, 148, 167, 184, 191
Philadelphia Museum of Art, 16n17
Ping!, 102–106
Plimpton, George, 72
pocket PC. *See* Personal Digital Assistant
podcast, xvii, 6, 63, 119
policy, 188
polling. *See* voting
popularity of digital exhibits, 38
portability, 37, 129, 133, 137
post-visit. *See* extending the visit
Powerhouse Museum, 16n17
presence, 131. *See also* proximity; sensors
pre-visit. *See* extending the visit
Presidio, 128
privacy, 139, 159
prototype, 46, 52, 125; paper, *136*
proximity, 131, 168
punch cards, xxiv, 127

radio, xxiv–xxv, 128
Radio Frequency Identification. *See* RFID
Really Simple Syndication (RSS), 11
reflection, 142. *See also* extending the visit
Reina Sofia, 82
Rembrandt, 80
research: eye-tracking, 67; focus groups, 135, 140; observation, 12, 39, 40, 43, 46, 66, 83, 99, 119, 142, 154–56, 160, 168; proprietary, xx, 67; psychological, 120; semiotics,

168; sociocultural, 168; summative, 171; surveys, 68, 70, 154, 171
resolution. *See* screen resolution
RFID, 85–86, 125, 130, 135–39
Royal Institution, 83
ruggedized hardware, 86

San Francisco Bay Area, 132
San Francisco Museum of Modern Art (SFMOMA), xiv, xvii, 5, 7, 12
San Jose Museum of Art, 10
Sandberg, Willem, x
scenarios, 134
schools, 111, 116, 120, 142, 156–61. *See also* teachers; visitors, young
science centers. *See* museum, science
science experiments, 126
Science Museum (London), 36, 39, 40, 44, 45, 118
Science Museum of Minnesota, 38
screen resolution, 87, 90
sculpture, 71–73, 75, 128
sensors, 127, 137, 168. *See also* infrared; location tracking; proximity; RFID; ultrasound
sensory information, 28, 81, 98, 101–106, 190; tactile model, 103
Sharples, Mike, 111
Short Messaging Service. *See* SMS
Short-Wave Ambulatory Lectures, xiii
Siemens, George, 11, 13
simulations, 141–43
smart cards, 137. *See also* RFID
Smithsonian Museum of Natural History, xxiv
Smithsonian Photography Initiative, 16n17

SMS, 86, 114
social construction, 112, 149
social interaction, 47, 52, 142, 184
social networking, 12, 164
Sony Walkman, xiv
speaker, 48. *See also* earpiece; headphones; headset
standardization, xxi, 188
"*Star Wars*: When Science Meets Imagination" exhibition, 84, 106–107
Stedelijk Museum, x, xiii, xvii, xxii
steve project, 8
storytelling. *See* narrative
structure, 64, 75–76, 109–21, 185. *See also* choice and control
Student. *See* schools; visitors, young
Study Gallery, 159

tablet PC, 125, 127, 151
tagging, 8, 11, 43, 114–15, 137. *See also* bookmarking
Tate Britain, 80, 120
Tate Modern, xvii, 7, 39, 41, 44, 46, 80, 81, 82, 83, *84*, 87, 90, 112
teachers, 114, 133, 157. *See also* classroom; schools
television, 10, 186
text, ix–xii, 180, 189
text messaging. *See* SMS
3D displays, 164
time, 74, 100, 111, 153–56, 159–60
tourists, 111
touchscreen, 43, 46, 79, 86
toys, 167–70
trails, 83, 109–21, 127, 148, 180
"Treasures of Tutankhamun" exhibition, xiv, xvii, xx, xxii
trompe l'oeil, 99

ultrasound, 154
Universeum Science Discover
   Center, 167, 170–72
University of Birmingham, 153
University of Glasgow, 150
University of Washington, 137
Urbis, 159
usability, 87
user-generated content, 141, 186–87
user model, 85

Vasari, Giorgio, xix
Versailles, 82
Victoria and Albert Museum (V&A),
   36, 38, 40, 110, 118–19
video, x, 10–12, 15, 36, 38, 41, 79,
   81–82, 86, 87, 163, 164. See also
   labels; video
"virtuous circle," 83–84, 89
visitors: blind, 101–106; deaf, 82,
   106–107; expectations, 155;
   groups of, 19; older, 100–101,
   144; segmentation of, 25, 51;
   young, 82, 99–100, 112, 114, 119,
   142, 144, 168–72. See also docent,
   tours; families; mental model;
   schools; tourists

voice: digital, x; messages, 149;
   museum, x, 5; multiple, 5. See also
   audio recording
voting, 86

Walkman. See Sony Walkman
Watch Water Freeze exhibit, 137
"The Water's Way" exhibition,
   170–72
Web, 50, 79, 88, 111–12, 118–20,
   133, 135, 137, 147, 150–53,
   158–62, 164, 183, 187, 189
wetlands, 127
Whitney Museum of American Art,
   66, 67, 70–76
WiFi, 85
Wikipedia, 149
Winckleman, J.J., xix
wireless, 85–86, 125, 132, 141. See
   also barcode scanning; Bluetooth;
   cell phone; GPS; infrared;
   network; RFID; WiFi
YouTube, 164

de Zeeuw, Gerard, xi
zoo, 41; London, xiv

About the Contributors

Jonathan P. Bowen is emeritus professor at London South Bank University and chair of Museophile Limited, a museum consultancy company. He is also a visiting professor at King's College London and a visiting academic at University College London. He has been involved with the field of computing in both industry and academia since 1977. As well as computer science, his interests extend to online museums and the history of computing. He was honorary chair at the first Museums and the Web conference in 1997 and has contributed to each conference since then. Bowen is a fellow of the Royal Society for the Arts and of the British Computer Society. He holds an MA degree in engineering science from Oxford University.

James M. Bradburne is director general at the Fondazione Palazzo Strozzi in Florence, Italy. He is a British-Canadian architect, designer, and museum specialist who has designed World's Fair pavilions, science centers, and international art exhibitions. Educated in Canada and England, he developed numerous exhibitions, research projects, and symposia for UNESCO, UNICEF, national governments, private foundations, and museums worldwide during the course of the past twenty years. Prior to heading Palazzo Strozzi, he served as head of Design, Research and Development at newMetropolis Science and Technology Center in Amsterdam and general director of the Applied Arts Museum in Frankfurt. He is also a research fellow at the

London Knowledge Lab and director of the Next Generation Foundation—an independent foundation to promote innovation in informal learning initiated by the president of LEGO, Kjeld Kirk Kristiansen. Bradburne sits on several international advisory committees and museum boards, has curated and designed exhibitions, lectures internationally about new approaches to informal learning, and has published extensively.

Alexandra Burch is manager of Audience Research and Advocacy in the Science Museum, London. She has worked at the Science Museum for seven years. Prior to this she graduated with a PhD in marine biology, having spent four years studying marine life in Wales. Since joining the Audience Research and Advocacy department, Burch has been involved in a number of different and innovative projects—she developed an intellectual framework for evaluating dialogue events, conducted research on understanding how visitors engage with content through websites, and worked on numerous exhibition projects. Burch's work focuses on ways to embed learning in the visitor experience, and her most recent work has concentrated on implementing novel methods of interpretation to help support social interaction.

Lynn D. Dierking is Sea Grant Professor in Free-Choice Learning at Oregon State University and senior researcher at the Institute for Learning Innovation. Dierking is internationally recognized for her research on the behavior and learning of children and families in free-choice learning settings and the development and evaluation of community-based efforts, and has published extensively in these areas. She received her PhD in science education from the University of Florida, Gainesville, and has worked in a variety of educational settings, including museums, schools, community-based organizations, and universities. She serves on the editorial boards of *Science Education* and the *Journal of Museum Management and Curatorship* and has been an advisor for several National Research Council efforts.

John H. Falk is Sea Grant Free-Choice Science Learning Professor, Oregon State University, and president emeritus of the Institute for Learning Innovation, an Annapolis, Maryland-based nonprofit learning research and development organization. He is known internationally for his research on free-choice learning—the learning that occurs in leisure settings like museums, parks, and on the Internet. Falk has authored over one hundred scholarly articles and chapters in the areas of learning, biology, and education as well as more than a dozen books and has helped to create several nationally important out-of-school educational curricula. Falk received a joint doctorate in biology and education from the University of California at Berkeley. He also earned MA and BA degrees in zoology and a secondary teaching credential in biology and chemistry from the same institution.

Silvia Filippini-Fantoni is completing her PhD at Sorbonne University on the use of personalization technologies in museums and is a visiting research fellow at London South Bank University. She has presented the results of her research at international conferences and seminars and has worked on projects with the Louvre and the City of Science and Industry in France, the Indianapolis Museum of Art, the Metropolitan Museum of Art, Tate Modern, and the J. Paul Getty Museum. More recently, Filippini-Fantoni was employed as an evaluator and product manger for multimedia at Antenna Audio, where she was responsible for the development of software and hardware applications for multimedia.

Ben Gammon started his career at the Science Museum, London, gaining experience in the curatorial, education, and traveling exhibition departments before leading the Audience Research and Advocacy department for seven years, and then working as head of Learning and Audience Development. Gammon is now an independent consultant specializing in audience research, interpretative planning, and training.

Ellen Giusti is a visitor studies consultant. She has worked with media developers for informal educational settings for almost twenty years. She spent fifteen years as in-house evaluator at the American Museum of Natural History (AMNH) where she worked on its first generation of interactive computer exhibits, designed for "Global Warming: Understanding the Forecast," which were cutting edge for their time. At the AMNH she also worked on all phases of audience research for AMNH's three audio guides and on measuring audience impact of interactive exhibits within informal science education. As evaluation consultant for Touch Graphics, Giusti continues to work on National Science Foundation (NSF)-funded projects that use cell phone technology to make exhibitions accessible to visitors who are blind or have low vision. She is a past chair of American Association of Museums' (AAM) Committee on Audience Research and Evaluation (CARE) and a board member of the Visitor Studies Association for six years.

Halina Gottlieb, art historian and multimedia producer, has been a program manager at Visions for Museums at the Interactive Institute (Stockholm) since 1999. As project manager she has taken part in the development of several prototypes pertaining to the interpretation of objects at art galleries. She is also curator of the Interactive Salon, a showroom for technologies that promote and preserve cultural heritage. In 2002, Gottlieb founded the conference/award forum Nordic Digital Excellence in Museums and Heritage Sites (NODEM), and in 2006 she became director of the Swedish Forum for Cultural Heritage at the Interactive Institute.

Sherry Hsi conducts research and evaluation for the Center for Learning and Teaching at the Exploratorium, San Francisco, in areas of new media projects, handhelds, science websites, digital libraries, and other informal science learning programs. Her research focuses on understanding how to design and evaluate social contexts for better science learning, facilitation, and deeper reflection mediated by new media, online environments, and networked technologies. She was awarded grants from the NSF National Science Digital Library program to create an exhibit-based learning and teaching digital library for use in after-school settings, and for educator professional development. She was also awarded a MacArthur Foundation grant to explore digital-mediated learning among next-generation youth. In 1996, she helped develop online teacher professional development for the first Virtual High School. She is coauthor of the book *Computers, Teachers, Peers: Science Learning Partners* (2000). Hsi is on the editorial board for the *International Journal of Science Education* and reviews for the *Journal of the Learning Sciences.*

Peter Lonsdale is a PhD research student at the University of Nottingham's Learning Sciences Research Institute. He has worked as a developer and consultant on a range of e-learning projects, including the EU-funded MOBIlearn project exploring next generation mobile learning systems and the MyArtSpace project looking at the use of mobile phones for learning in museums. He also worked as an e-learning developer at the National College for School Leadership in Nottingham, in a team responsible for maintaining and supporting its online learning management system. Lonsdale has a master's degree in the design of intelligent systems and human-computer interaction, and his PhD research centers on the development and evaluation of software to allow children to learn in physical spaces, using location-based activities running on handheld computers.

Julia Meek is an independent evaluation consultant specializing in evaluating the use of information and communications technology in learning. Meek has over ten years' experience of developing computer-based learning material for higher education and supporting academics to develop, integrate, and evaluate the use of learning technology in their teaching. Her PhD research at the University of Birmingham focused upon the design and development of the "Evaluation Lifecycle Toolkit," which now forms the core business of her LIFECYCLE consultancy. Her evaluation projects include Training Researchers in Online Research Methods, a project funded by the UK's Researcher Development Initiative, running from 2007 to 2009, at the University of Leicester. It is a comprehensive, three-stranded training program in online research methods. She also evaluates Managing Research Projects, running from 2007 to 2008, at the University of Warwick. The goal of this project is to harvest the wealth of project management experience and expertise gained by social science researchers in higher education.

Ross Parry is lecturer in museums and new media in the Department of Museum Studies at the University of Leicester and was made an Innovations Fellow by Higher Education Innovation and Regional Fellowships for his work developing in-gallery digital media. As a historian of museum media, his research explores the ways memory institutions use the new technologies of their times (whether in a digital or predigital age) to manage information and display knowledge. He has published widely on the subject of museum computing and is author of the first major history of museum computing: *Recoding the Museum: Digital Heritage and the Technologies of Change* (2007).

Paul Rudman is a research fellow in the computer science department at Oxford Brookes University. He has a background in commercial software development, with a BS in psychology and a MA in human-centered computer systems. His doctoral research at the University of Birmingham created a novel form of dynamic support for learning during spoken conversations. Rudman has since worked as research fellow on the Equator project at the University of Glasgow (funded by the UK's Engineering and Physical Sciences Research Council), investigating the interactions between a physical city and its digital representation, and as consultant on the evaluation of MyArtSpace, a mobile learning service to support school museum visits. Other work includes the evaluation of the EU-funded AtGentive project, which investigates the role and effects of learner attention in a children's online learning environment. Rudman's research interests focus on the use of technology to support the learner and the learning process, especially with regard to informal and mobile learning.

Peter Samis is associate curator of interpretation at the San Francisco Museum of Modern Art (SFMOMA). In 1993, he served as art historian for the first CD-ROM on modern art and then spearheaded development of SFMOMA's Interactive Educational Technologies (IET) program. Since then, programs produced by the IET team have received wide recognition from sources as diverse as the American Association of Museums, the Webby awards, Communication Arts, and I.D. Magazine. Samis is on the governing councils of two museum-focused open source initiatives: Pachyderm 2.0 (www.pachyderm.org) and steve (www.steve.museum). Together with his team, he continues to produce innovative content for SFMOMA's galleries, website, podcasts, and the Koret Visitor Education Center.

Mike Sharples is professor of learning sciences and director of the Learning Sciences Research Institute at the University of Nottingham. He has an international reputation for research in the design of learning technologies. He inaugurated the mLearn conference series and served as deputy scientific manager of the EU-funded Kaleidoscope Network of Excellence in Technology Enhanced Learning. His current projects include a collaboration with the Open University to develop personal technology to support inquiry science learning between home and school. Recent projects include MyArtSpace for mobile learning in museums and the L-Mo project with Sharp Laboratories of Europe to develop handheld technologies for language learning. He is author of 160 publications in the areas of interactive systems design, artificial intelligence, and educational technology.

Jeffrey K. Smith holds a chaired professorship in the faculty of education at the University of Otago in New Zealand. Prior to coming to Otago, he was professor and chair of the Educational Psychology Department at Rutgers University in New Jersey, where he had been a faculty member for twenty-nine years. From 1988 through 2005, he also served as head of the Office of Research and Evaluation at the Metropolitan Museum of Art. He studies issues in educational assessment, learning in cultural institutions, and the psychology of aesthetics. He received his BA from Princeton University and his PhD from the University of Chicago. He is currently coeditor of the journal *Psychology of Aesthetics, Creativity, and the Arts*.

Loïc Tallon is an independent researcher specializing in digital media initiatives in the museum and the evolution of visitor-museum dialogue. He was educated at Nottingham University and the Courtauld Institute of Art where, at the latter, he concentrated on the history of the museum and display. His thesis was on the history of the audio tour; it was while writing his thesis that the vision for this book was conceived.

He has presented his research at Museum Association UK, American Association of Museums, and AVICOM conferences. Working as a project manager and digital media specialist with Cultural Innovations, a leading international museum consultancy, Tallon has gained international experience on major museum projects in the United States, Singapore, Syria, India, Egypt, France, and the UK.

Pablo P. L. Tinio is a junior research scientist in the Department of Psychology, University of Vienna. Tinio holds an MA in behavioral science from Kean University and an MA in education—learning, cognition, and development—from Rutgers University. He is presently pursuing a PhD in psychology at the University of Vienna, where his research is focused on the psychology of aesthetics, art, and creativity. He has previously conducted research in museums such as the Whitney Museum of American Art. Tinio's work is informed by his experience as a professional photographer and artist.

Giasemi Vavoula is an RCUK Academic Fellow at the Department of Museum Studies, University of Leicester, UK. She has a background in computer science (BS) and human-centered computer systems (MS), while her doctoral research at the University of Birmingham focused on the design of personal lifelong learning organizers. Vavoula has previously worked as a research fellow on the EU-funded MOBIlearn project and the Kaleidoscope Network of Excellence, and also as a consultant on the evaluation of MyArtSpace (a mobile learning service to support school museum visits) and the requirements analysis of the project Resources for Learning by Exploration (which used semantic Web technologies to support students' use of course materials and related online resources). Her research interests currently focus on technology-enhanced museum learning, tools and methods for mobile and informal learning research, and the long-term impact of school visits to science centers on students' learning about science. Vavoula is currently involved in the design of a new master's program at Leicester on digital heritage.

Kevin Walker has worked in and with museums since 1996, including five years as senior software designer for Exhibitions at the American Museum of Natural History in New York and several years as independent consultant and designer. His permanent and temporary museum installations and websites have won awards internationally and traveled worldwide. At the London Knowledge Lab he now works on projects in technology-enhanced learning and international development. He holds a BA in anthropology and mass communications from the University of California, Berkeley, and a MA in interactive telecommunications from New York University. His PhD research at the Institute of Education, University of London, has focused on personalized learning trails created by museum visitors using mobile technologies. He has published widely on education, technology, and museum issues, and writes a regular column for *Educational Technology* magazine.